SpringerBriefs in Economics

More information about this series at http://www.springer.com/series/8876

Jingjing Yan

Comprehensive Evaluation of Effective Biomass Resource Utilization and Optimal Environmental Policies

 Springer

Jingjing Yan
School of Humanities and Economic
   Management
China University of Geosciences
Beijing
China

ISSN 2191-5504   ISSN 2191-5512   (electronic)
ISBN 978-3-662-44453-5 ISBN 978-3-662-44454-2   (eBook)
DOI 10.1007/978-3-662-44454-2

Library of Congress Control Number: 2014946959

Springer Heidelberg New York Dordrecht London

Printed on acid-free paper

Springer is part of Springer Science+Business Media (www.springer.com)

Acknowledgments

I express my deep gratitude to Dr. Yoshiro Higano, Professor, Doctoral Program in Sustainable Environmental Studies, University of Tsukuba, for his careful guidance in preparing the content with intense concern, encouragement, and patience. I am grateful to him for his valuable suggestions and motivations during the study, from which I have greatly benefited. I am also grateful to Professor Zhenya Zhang, Professor Chuanping Feng, Professor Norio Sugiura, and Professor Suminori Tokunaga for their valuable comments and advice.

I am gratefully obliged to Professor Jinghua Sha, Professor Rongge Xiao, and Professor Jinsheng Zhou, China University of Geosciences and Mr. Runbo Gu for their constant inspiration, help, and strong encouragement during my study life.

Special thanks go to Professor Yalin Lei, Professor Haizhong An, China University of Geosciences, who kindly gave me constant support and encouragement throughout the entire research work.

I acknowledge my heartfelt thanks to the China Scholarship Council (CSC) and China University of Geosciences for awarding me the National Construction High Level University Government-sponsored Graduate Student Project, National Natural Science Foundation of China, the Fundamental Research Funds for the Central Universities, and Beijing Social Science Fund.

Lastly, but most of all, I am obliged to my husband Mr. Xiao Chu and my son Hanbin Chu who give constant deep love, support, and understanding for our happy and blissful future. I am especially obliged to my parents Mr. Jianrong Yan and Mrs. Zhongxiang Ren, who always provide me emotional support, patient understanding, and ceaseless care and love during my whole life. This book is dedicated to the memory of my mother, who gave me what common sense I have. I miss her deeply, though I know she will always be a part of us.

Supports

This research was supported by the National Natural Science Foundation of China (41101559), the Fundamental Research Funds for the Central Universities (2652012084) and Beijing Social Science Fund (14JGC116).

Contents

Figures

Tables

About the Author

Dr. Jingjing Yan, female, associate professor of economics at China University of Geosciences in Beijing. Acquired Doctoral degree in environment and resource industry economics, graduated from University of Tsukuba in Japan and China University of Geosciences, Beijing in China. Main research fields include resource and environment economics, regional economics, and industrial economics. As the head and coordinator of projects, she carried out a series of projects of Natural Science Foundation of China (NSFC) at the provincial and ministerial level, and wrote a series of articles about resource and environment evaluation, which was published in international journals.

Abstract

The stockbreeding industry is booming rapidly in rural areas around big cities in China, especially in the rural areas around Beijing, carrying high risk to the environment due to emission of huge amounts of pollutants such as COD, T-N, and T-P as well as greenhouse gases into rivers and atmosphere. On the other hand, as typical biomass resource, stockbreeding waste can be used as energy source by advanced technologies. In this research, we selected Miyun County of Beijing and focused on analyzing the synthetic environmental management policy by computer simulation including the introduction of two advanced technologies to improve the environment and provide more biomass energy. The model considers both the total ecological system in the objective region and the social-economic situational changes over the period. The purpose of our research is to establish effective utilization methods of biomass resource with coordination among resource reutilization, environmental preservation, and economic development, and finally realize the sustainable development of the society.

Miyun County is the largest rural suburb and the most important county of Beijing. Especially Miyun Reservoir, the only surface water resource of Beijing, provides more than 60 % of the water resources for Beijing City. This important ecological value of Miyun County has unique significance for the regional environment and dramatically influences the water safety and local development of Beijing. In the catchment areas of Miyun Reservoir, stockbreeding wastes are generated and heavily contribute to water pollution, especially pig farming has become the source of the most serious pollutants of Miyun Reservoir. On the other hand, stockbreeding waste is a kind of biomass resource of carbon neutral which can be utilized to produce biomass energy through advanced technologies. Therefore, we should consider the comprehensive utilization of livestock feces and urine as a biomass resource when we formulate synthetic policy to reduce water pollutants emitted from stockbreeding industry. In this paper, we also consider reduction of greenhouse gases as a whole ecological system based on material balance.

We focused on analyzing the synthetic environmental management policy by computer simulation including the introduction of advanced technologies to improve environment and provide more biomass energy. We considered both the ecological system in the objective area and the socioeconomic situational changes during a specific period. To achieve our research purposes, one objective function (minimize total nitrogen (T-N) over the simulation running period), the water pollutants flow balance model, the air pollutants flow balance model, the energy balance model, and the socioeconomic model were specified to express all key factors and parameters reflecting the environmental situation and affecting human activities; then the computer simulation models were used to analyze the synthetic policies with introduction of two advanced technologies in the objective region.

In this study, when we adopted the policy to introduce two advanced technologies (*New Energy Project* and *Biomass Energy Recycle Plant*), the policy was a very effective tool compared to the present stockbreeding wastes treatment technology to reduce environmental pollutants in all simulations. The introduction of two advanced technologies raised the level of economic growth by 10 % compared to not adopting the advanced technologies policy. When the two advanced technologies were introduced, the objective value (total T-N) showed a reduction of about 4,916 tons compared to not adopting the advanced technologies. Moreover, we also found that if the policy to introduce the advanced technologies was adopted in the catchment area, production of the pig farming industry increased to about 448 million RMB Yuan over the total 10-year simulation period compared with not adopting.

Concerning environmental pollutants, we found a reduction of about 168,000 tons of greenhouse gases emitted by the pig farming industry in the total 10-year simulation (2007–2016) when we introduced the advanced technologies. In addition, we found that about 6,424 tons of T-N, 2,716 tons of T-P, and 42,729 tons of COD emitted by the pig farming industry in the total 10-year simulation can be reduced by introducing the advanced technologies. From the aspect of biomass electric energy, more than 32.7 million kWh of electric energy can be generated when we adopt the two advanced technologies.

The results of this study establish the synthetic policies for the catchment area, especially the introduction of advanced technologies for the pig farming industry are very effective and the utilization of biomass resources allows simultaneous pursuit of environmental preservation and economic development. This comprehensive evaluation is expected to improve the biomass resource utilization and environmental reservation and form the basis of policy decision-making for sustainable development of rural areas of big cities in China.

Chapter 1
Introduction

Abstract The stockbreeding industry is booming rapidly in rural areas around big cities in China, especially in the rural areas around Beijing, carrying high risk to the environment due to emission of huge amounts of pollutants, such as COD, T-N, and T-P, as well as greenhouse gases into rivers and atmosphere. On the other hand, as typical biomass resource, stockbreeding waste can be used as energy source by advanced technologies. The purpose of our research is to establish effective utilization methods of biomass resource with coordination among resource reutilization, environmental preservation, and economic development, and finally realize the sustainable development of the society. The background, objectives, and methods of this book are introduced, and all the concepts and literature reviews are shown in this chapter.

Keywords Biomass resource · Stockbreeding industry · Rural areas

1.1 Background

The rapid industrialization in developing countries like China, though contributed to economic development has resulted in heavy losses to environmental degradation in terms of effects on human health and ecosystem at large through air and water pollution. Basically water pollution poses a serious challenge due to its impact on a large number of socio-economic activities. Specially, contamination of surface water resources in rural areas is seen by Chinese government as the nation's most critical environmental priority. Ever since 1996, China has designated water pollution prevention as the essential work of environmental protection of the entire country. Most local governments of rural areas have implemented policies with adopting wet methane fermentation technology. However, according to the report of the state environmental protection administration of China, still more than 50 % of surface water resources in rural areas do not conform to the national drinking water standards (State Environmental Protection Administration 2008).

© The Author(s) 2015
J. Yan, *Comprehensive Evaluation of Effective Biomass Resource Utilization and Optimal Environmental Policies*, SpringerBriefs in Economics,
DOI 10.1007/978-3-662-44454-2_1

In rural areas around big cities in China, with the improvement in living standards, demand for animal products dramatically increases. The stockbreeding industry is expanding rapidly in rural areas around large Chinese cities such as Beijing. The gross output value of stockbreeding industry accounts more than half percent of the total agricultural output value in rural areas. According to the report of State Environmental Protection Administration, till 2010, the growth rate of gross output value of stockbreeding industry will keep increase ratio as 8 % per year (State Environmental Protection Administration 2002). However, majority of stockbreeding wastes have been improperly treated and carry a high environmental risk in terms of COD (chemical oxygen demand), T-N (total nitrogen), T-P (total phosphorus) and GHG (green house gases) levels because of large amounts of pollutants and greenhouse gases released to rivers and the atmosphere. Now, stockbreeding wastes have become the most serious pollution source in rural areas. The annual gross output of stockbreeding waste is about 1.73 billion ton, which is 2.7 times of the industrial waste (State Environmental Protection Administration 2002). Especially in the big cities and provinces like Beijing, Shanghai, Henan Province, Zhejiang Province and Guangdong Province, the stockbreeding waste have serious impaction on the environment: (1) there are more than 80 % of the large-scale farms without the necessary facilities to treat the concentration of water pollution; (2) about 25 to 30 % of total net load of water pollution COD (chemical oxygen demand), T-N (total nitrogen)and T-P (total phosphorus) emitted by stockbreeding wastes directly flowed into surface water in rural areas; (3) according to the survey of environmental pollution in Beijing, the quantity of NH_3 (ammonia) emitted by stockbreeding wastes is about 24,330 ton per year, which account for 34 % of total ammonia emissions of Beijing (State Environmental Protection Administration 2002).

However, as a typical biomass resource of carbon neutral, stockbreeding wastes can be used as energy source in advanced technologies. Through the main biomass technologies (e.g. methane fermentation), it is good not only in the sense that it provide energy and materials of carbon neutral and it provides new business chance but also in that the utilization itself is one of proper and perhaps best treatment way of wastes in stockbreeding industry. Biomass conversion and energy utilization technology offers a very significant potential for increasing energy production and reducing pollution.

Therefore, it is absolutely of necessity and possibility to research the optimal policies and technologies for effective utilization of stockbreeding wastes as biomass resource and improvement of environment in rural areas around big cities in China. Beside these, with the integrated policies, we can also provide the biomass resource and energy (biogas and fertilizer) and realize the harmonious development of economy, resource and environment, which would be feasible and foreseeable in the future. Therefore, in formulating effective policies to reduce the quantities of water pollutants and green house gases (GHGs) in rural areas, we need to construct a model that describes socio-economic activities, the load of pollution and imply instruments in order to reduce pollution without deteriorating the socio-economic development.

1.2 Statement of the Problem

In the circumstances clarified above, finding solutions to improve the environmental status and realize effective utilization of stockbreeding wastes are the critical issues. In this study, we selected the upstream area of Miyun Reservoir in Miyun County, Beijing, as the study area, and efforts have been made to formulate effective policies to utilize stockbreeding wastes as biomass resource by advanced technologies and reduce the quantities of water pollutants and green house gases (GHGs) in this area, and construct a model that describes socio-economic activities, the load of pollution and imply instruments in order to reduce pollution without deteriorating the socio-economic development.

The main part of this study is the integration of the ecosystem, as water quality and green house gases indicators related to macroeconomic structure, and indicators into mathematical quotations. The framework is actually based on a conceptual model, which sets the links between ecosystem processes, functions and outputs of goods and services with dynamic simulations of processes in the system. These can be used to explore the economic development, environmental preservation and effective biomass resource utilization.

In addition, after the basic simulation for current situation, the advanced technologies and integrated policies are introduced to find and optimal solution set with some policy measures in formulating a balance between economic development and environmental improvement as water quality objectives. Moreover, the integrated modeling discusses applying the rural areas around big cities as a comprehensive approach to address the problem. Finally, a conclusion with policy proposals that would be possible to provide appropriate action is introduced (see Fig. 1.1).

The content of this book consists of seven chapters. Chapter 1 introduces the background, contents, objectives, methods and literature reviews. Chapter 2 is a descriptive analysis of present environmental condition in Beijing and the pollution emitted by stockbreeding industry in the study area. It gives some details on water contamination problem with respects to stockbreeding wastes and the reason why we select the upstream area of Miyun Reservoir in Miyun County, Beijing as the study area. Chapter 3 analyzes and simulates the present environmental policies with adoption of current technology (wet methane fermentation) in the study area, in order to realize the water pollution minimization and economic development. Based on the results of the basic simulation in Chaps. 3 and 4 represents the concept of optimal environmental policies and reconstructs the integrated policy framework with the introduction of advanced technologies for proper treatment of stockbreeding wastes. Chapter 5 constructs the comprehensive evaluation model to simulate the integrated policies and evaluates whether the integrated policies with introduction of advanced technologies can reach the purpose of simultaneous pursuit of environmental preservation, effective utilization of biomass resource and economic development. Finally, Chap. 7 concludes the thesis by summarizing the major findings and proposing further research ideas.

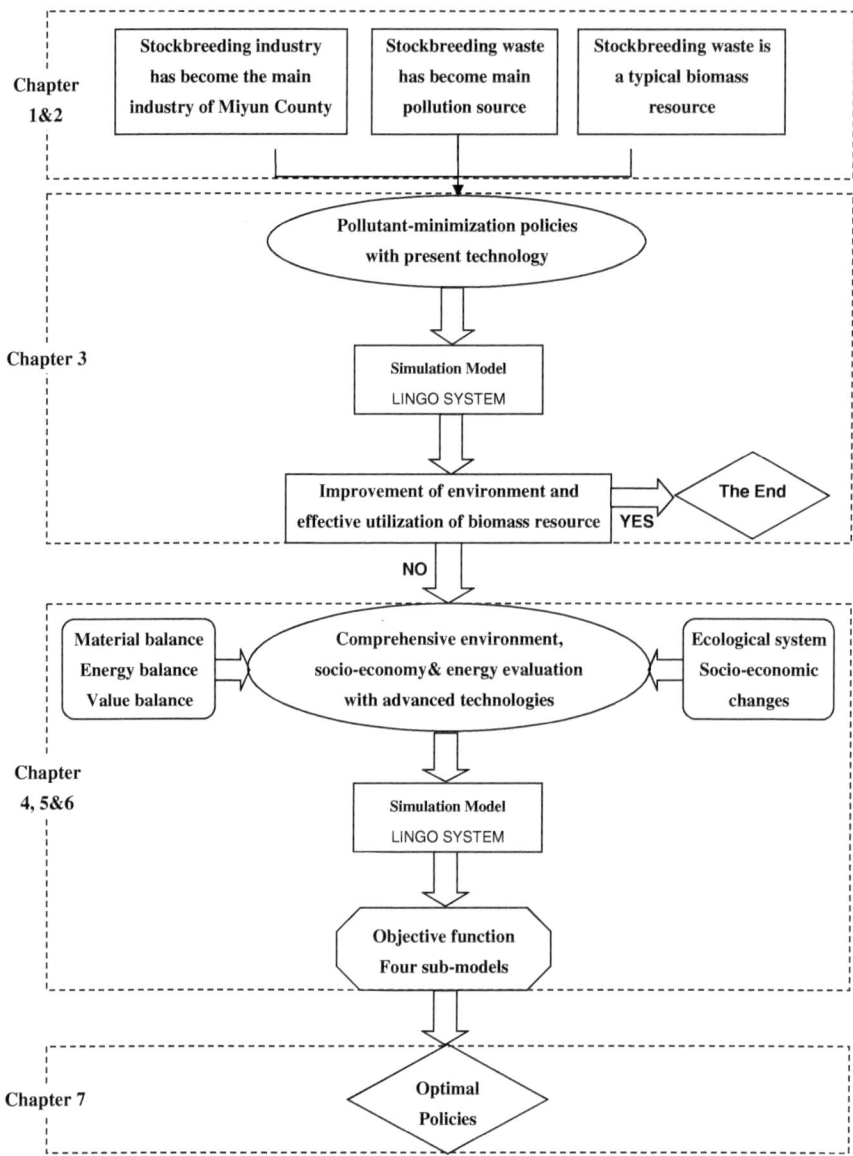

Fig. 1.1 Framework of the research

1.3 Objectives of the Study

The specific objectives of this research are: (1) to analyze the problem from the view of economy, environment and resources, and it will establish theoretical system of the biomass resource utilization, which including the evaluation of technology, environment, resource and economy, and this comprehensive assessment would be scientific and practical; (2) to develop a dynamic mathematical model in terms of China's economic characteristics and simulate the interrelation between the environment and socio-economic structure in the study area, in order to observe if the general wet methane fermentation technology can satisfy the requirements of environment and economy; (3) to introduce advanced technologies from Japan and China; and reach a decision regarding the optimal solution to the comprehensive problem by introducing advanced technologies and realize simultaneously pursuit of environmental preservation, effective utilization of biomass resource and economic development.

1.4   Methodology

We formulate the comprehensive system as a linear optimization model which we solve a mathematical optimization software package, LINGO. This enables us to simultaneously simulate the water environmental system and the socio-economic system by creating as much linearity as possible in the functions that is close to the reality. Detailed discussions of linear programming theory may be found in books by Hadley (1962), Hillier and Gerald (1967) and Wagner (1959).

Computable models are common methods that have been used in quantifying the impact of environmental problems. They are usually utilized to give predictions, which can be employed in the process of decision-making. They also can be evaluated quantitatively against alternative models. Because we can only observe limited aspects of a system, environmental or economic, computer modeling is the recognized vehicle for application of the scientific method in studying the dynamics of environmental and economic systems. Despite of its imperfection, it continues to yield improvements in comprehension and prediction. Without modeling, it would be difficult to measure progress in understanding and to generate and evaluate new scientific insights.

The principle method used in this research is to develop conceptual mathematical models based on several theoretical principles. These models employ both secondary and primary data. The first model is a dynamic mathematical model. It is built based on the principal of material balance, Harrod-Domar model, input output table and other macroeconomic indicators to analyze the interrelation between water environmental system and socio-economic system and evaluate efficiency of the integrated pollution-minimization policies with adoption of wet methane fermentation technology. Second, an integrated mathematical model from environment, energy and socio-economic systems is developed to examine a set of policies with introduction of advanced technologies to reduce environmental pollution and improve regional development in the study area.

The data were constructed using census data, official data from governmental sources and the finding of previous studies.

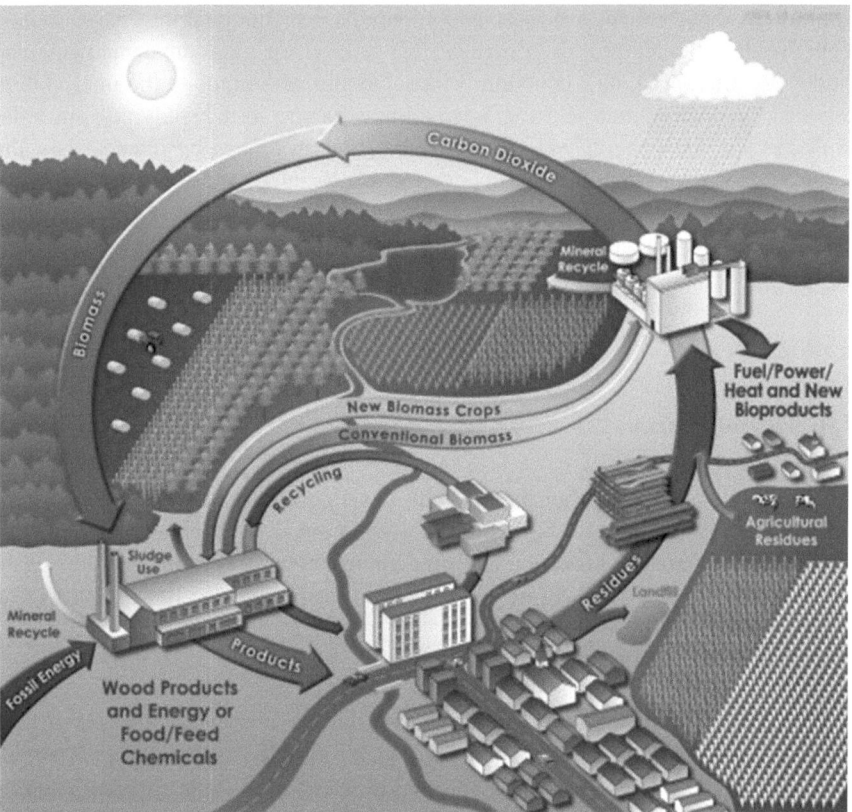

Fig. 1.2 Cycle of biomass resource *source* http://www.bioenergy.ornl.gov/papers/misc/bioenergy_
cycle.html, Dec 21, 2008

1.5 Definitions of Concepts

1.5.1 Biomass Resource

Biomass, a renewable energy source, is biological material derived from living, or recently living organisms, such as wood, waste, and alcohol fuels. Biomass is commonly plant matter grown to generate electricity or produce heat. For example, forest residues (such as dead trees, branches and tree stumps), yard clippings and wood chips may be used as biomass. However, biomass also includes plant or animal matter used for production of fibers or chemicals. Biomass may also include biodegradable wastes that can be burnt as fuel. It excludes organic material such as fossil fuel which has been transformed by geological processes into substances such as coal or petroleum (see Fig. 1.2).

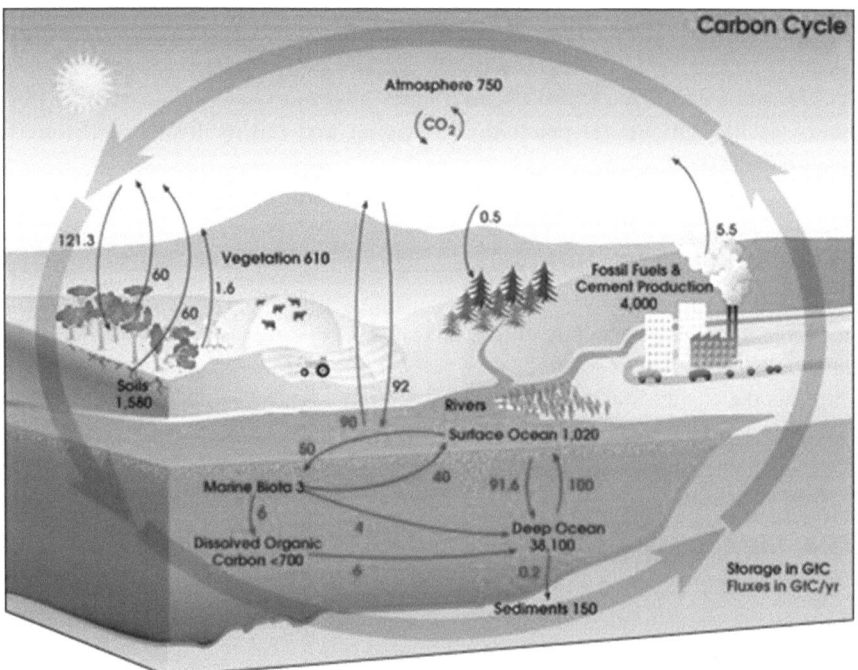

Fig. 1.3 Carbon cycles of biomass resource *source* http://www.rst.gsfc.nasa.gov/Sect16/carbon_cycle_diagram.jpg, Dec 21, 2008

Although fossil fuels have their origin in ancient biomass, they are not considered biomass by the generally accepted definition because they contain carbon that has been "out" of the carbon cycle for a very long time. Their combustion therefore disturbs the carbon dioxide content in the atmosphere.

Biomass resource as a renewable and clean energy, it has many merits than other resources. Though the technology of biomass resource, it not only can provide energy and materials of carbon neutral and new business chance, but also provide proper best treatment way of stockbreeding wastes.

There are countless benefits of using biomass. They include:

- Economic Benefits. Biomass could replace some of the money spent on oil.
- Environmental Benefits: (1) Preservation of agricultural land that otherwise would be sold for residential development or industrial use; (2) Sustainable agricultural techniques for these crops can restore and ensure soil stability and health and minimize chemical residues and habitat destruction; (3) Methane is 20 times more powerful than CO_2. Capturing methane from producers such as cows or rice fields and using it as fuel will significantly reduce this greenhouse gas; (4) Increased carbon sequestering from the crops grown for biomass; (5) Use of waste from agricultural and timber industries; (6) No net increase in atmospheric carbon dioxide (see Fig. 1.3).

1.5.2 Stockbreeding Industry

Stockbreeding industry is also called animal husbandry, animal science or husbandry, is the agricultural practice of breeding and raising livestock (Biomass Handbook 2003).

1.5.3 Stockbreeding Wastes

Stockbreeding wastes generate during feces, sewage, dust, odor, noise, which have impactions on rural atmosphere, water, soil and the cross-composite influence.

In this paper the stockbreeding waste specially refer to the feces and urine and pollutants caused by the stockbreeding industry (Biomass Handbook 2003).

1.5.4 Big City

According to the definition of City Economics, the population over 500,000 is big city in China.

1.5.5 Present Technology

In this paper, we named the general wet methane fermentation technology as present technology (this technology has been widely used in the study area to treat the wastes of stockbreeding industry).

1.5.6 Advanced Technologies

In this paper, we called two technologies (*New Energy Project & Biomass Resource Recycle Plant*) as advanced technologies because of the high efficiency in reduction of environmental pollutants and production of biomass energy as compared to other general stockbreeding treatment methods.

1.6 Theoretical Background

We have developed an ecological-economic model by integrating an input-output model of an economy with common water pollution parameter such as COD, T-N and T-P in the upstream area of Miyun Reservoir in Miyun County using economic and ecological data from the region. A model of this type can be represented as a system of linear equations and the principles of linear algebra can be used to calculate the values of variables such as sector's output or amount of pollution (Jin et al. 2003).

Theoretically, the integrated modeling approach from the points of studies that support developing modeling in this study summarized in the following:

Integrated approach was used in modeling since the late 1960s, which input-output models extended linking to environmental or natural resource area. Most of them have been concerned with the effects of pollution from one or more industrial sectors on the output of other sectors such as Ayres and Kneese (1969), Leontief (1977), Forsund (1985), Lee (1982) and Perrings (1987). In many integrated economic-ecological models, natural resource inputs and pollutants are expressed in physical units, while economic exchanges are expressed in monetary units (Hannon 2001; Leontief 1977).

In corporation of linkage between environmental change and socio-economic development can not be attained by observational studies alone. Modeling of key environmental and socio-economic processes is an essential tool, required to strengthen management institutions and practice. The framework is itself based on a conceptual model, which sets the links between ecosystem processes, functions, and outputs of goods and services with dynamic simulations of processes in the system, which can be used to explore the consequences of environmental damage, and produce forecasts of future changes (Turner 2000).

Integration of the two systems initially needs to be formulated. More specifically, the formulation should result in a definition of the problem to be addressed, an identification of system limitations and constraints and recognition of the policy issues and objectives and the ways in which they can be evaluated. The objective in this approach is to clarify multi-sectoral interrelationships and to highlight the dynamic characteristics of ecosystem and socio-economic changes. Thus, activity models are the ways in which socio-economic drivers and pressures are related to pollutants fluxes in drainage basic networks (Turner 2000).

An integrated modeling, several aspects are significant to be concerned such as: how environmental policy will affect macro variables for examples; the level of output in an economy or GDP. Secondly, how impacts are distributed across different sectors of the economy and how changes in policy, for example, impact on the long-term growth rate of the economy (Moffatt and Hanley 2001).

The modeling of the two systems, I-O table can be usefully applied up to the scale of a country, which could predict residuals of nutrients, sediments and other substances for a geographical set of economic activities and population settlements under a number of different economic growth scenarios expressed by Turner (2000).

Mathematically, the way of estimations methods and the description of the variables in equations are used based on the studies carried out in the Lake Kasumigaura by Higano and Sawada (1997), Higano and Yoneta (1999) and Hirose and Higano (2000). These studies address the lake pollution issues considering socio-economic activities related to reduction of pollution with introducing integrated dynamic linear or non-linear modeling approach.

1.7 Literature Review

Present evaluation theories and empirical researches of comprehensive evaluation are mainly focused on developed counties. Baumol and Oates (1988) and Oka (1997) analyzed the political aspects of the environmental problems. Arrow (1995) researched the carrying capacity of economic growth and environment. In Japan, Higano and Yoneta (1999) and Takagi (1999) constructed simple simulation analyses to evaluate water purification policies. In the recent years, many studies have explored how to protect the water environment and improve economic development. Hirose and Higano (2000) constructed simulation analyses to integrated environmental policies in the catchment area of Lake Kasumigaura, Japan. Subsequently, Mizunoya et al. (2006) extended and assessed synthetic environmental policies to reduce environmental burdens by biomass technology.

1.7.1 System Research on Biomass Resource

Biomass resource determination system studies the influencing factors of biomass resource industry development and its mechanism. The significance of determination system is the theoretical support for the government (see Table 1.1).

Mitchell (2000) evaluated the biomass DSS decision support system and suggested to construct a synthetic model with the advantages of different DSS models. But she did not analyze how to build the synthetic model based on the characteristics of different kind of biomass resources.

Hektor (2000) compiled the past decades on the utilization of biomass, divided scheme into national plans, projects plans and management plans. But Hektor did not make empirical research and analyze the specific economic measures when the administrative method failure.

Gielen et al. (2003) analyzed the impaction on the development of biomass energy with carbon tax. But he did not study the criteria of the carbon tax.

The Japanese energy association (2003) published *Biomass handbook* to introduce the biomass resource and the development, the book introduced the basic acknowledge of biomass resource and the advanced technologies, but it did not analyze how to evaluate the effects of biomass utilization and the advanced technologies.

1.7.2 Policy Research on Biomass Resource

Hillring (1998) analyzed the role of the government based on the review of different biomass energy strategies, emphasized the importance of technology development and the economic methods, than concluded the proper policies of government. But Hillring did not analyze how to utilize economic policies specifically.

Table 1.1 Theoretical research on biomass resource

Research	Author	Year	Achievements	Limitation
System Research	Mitchell	2000	Analyzed biomass DSS decision support system and suggested to construct a synthetic model with the advantages of different DSS models	Did not analyze how to build the synthetic model based on the characteristics of different kind of biomass resources
	Hektor	2000	Compiled the past decades on the utilization of biomass, divided scheme into national plans, projects plans and management plans	Did not make empirical research and analyze the specific economic measures when the administrative method failure
	Gielen et al.	2003	Analyzed the impaction on the development of biomass energy with carbon tax	Did not study the criteria of the carbon tax
	Japanese energy association *Biomass handbook*	2003	Introduce the biomass resource and the development, the book introduced the basic acknowledge of biomass resource and the advanced technologies	Did not analyze how to evaluate the effects of biomass utilization and the advanced technologies
Policy Research	Hillring	1998	Analyzed the role of the government based on the review of different biomass energy strategies, emphasized the importance of technology development and the economic methods	Did not analyze how to utilize economic policies specifically
	Coelho and Bolognini	1999	Summarized the non-technical hamper factors of the utilization of biomass energy, and held that the externality is an important influencing factor of biomass price	Did not construct the quantitative model between the price and the development of biomass resource
	Ray	2000	Utilized POLYSYS model measured potential allowance of biomass energy, pointed that it is essential to connect the agricultural with environment, regional economy and related industry	Did not calculate the correlation degree among the important factors

Coelho and Bolognini (1999) summarized the hamper of the utilization of biomass energy, in particular non-technical factors. The author held that the externality is an important influencing factor of the biomass resource price. But he did not construct the quantitative model between the price and the development of biomass resource. Ray (2000) utilized the POLYSYS model in order to analyze the impaction of agricultural policies on biomass energy. Ray measured the potential allowance of biomass energy and biodiesel, and pointed that it is essential to connect the agricultural with environment, regional economy and related industry, so as to improve the development of biomass industry. But the study did not calculate the correlation degree among the important factors.

1.7.3 Empirical Qualitative Analysis on Biomass Resource

Till now, there are lot of researches studied the biomass utilization situation, and majorities of them are qualitative analysis. The representative researches are listed here. Seungdo and Bruce (1999) studies the experience of biomass utilization in Finland, it concluded that biomass resource of forestry play a very important role in the biomass industry development, and biomass utilization would make contribution to meet the discharge standard of Kyoto Protocol. Ramachandra et al. (2000) analyzed the development of biomass energy in the UK, and proposed that it is important to strengthen the corporation between surplus region and scarcity region of biomass resource. Walsh et al. (2003) researched the development of biomass energy in the United States. Hillring (1998) introduced the biomass utilization in Sweden. Mandal et al. (2002) studied the biomass resource of rural areas of Indian, and analyzed the energy and efficiency of production system in the rural areas. In summary, previous qualitative studies of biomass resource utilization mainly focus on country and national scales. Little attention has been paid to the biomass utilization of cities or rural areas, and majority of the researches did not analyze the biomass utilization from the synthetic viewpoint of economy, society, resource energy and environment in the region, and construct the comprehensive model to do the research systematically (see Table 1.2).

1.7.4 Empirical Synthetic Analysis on Biomass Resource

In the field of synthetic evaluation, it was represented by Higano et al. (2006), who evaluated biomass resource from both the total ecological system and the socio-economic situational changes over a certain period, proposed a new aspect of synthetic environmental evaluation and policy to control the pollution, to realize the optimization and harmonious of the society. And the research of Higano is the main reference and investigative basis of this paper. Jolliet and Goedkoop (2004) adopted LCA (Life cycle assessment) as the method to analysis the biomass

Table 1.2 Empirical research on biomass resource

Research	Author	Year	Achievements	Limitation
Qualitative analysis	Seungdo and Bruce	1999	Studies the experience of biomass utilization in Finland	Little attention has been paid to the biomass utilization of cities or rural areas, and majority of the researches did not analyze the biomass utilization from the synthetic viewpoint of economy, society, resource energy and environment in the region, and construct the comprehensive model to do the research systematically
	Ramachandra et al.	2000	Analyzed the development of biomass energy in the UK	
	Walsh et al.	2003	Researched the development of biomass energy in the United States	
	Hillring	1998	Introduced the biomass utilization in Sweden	
	Mandal et al.	2002	Studied the biomass resource of rural areas of Indian, and analyzed the energy and efficiency of production system in the rural areas	
Synthetic evaluation	Higano et al.	2006	Evaluated biomass resource from total ecological system and the socio-economic situational changes, proposed a new aspect of synthetic environmental evaluation and policy to control the pollution, to realize the optimization and harmonious of the society	The model did not carried out with empirical application analysis by advanced biomass technologies
	Jolliet and Goedkoop	2004	Adopted LCA as the method to analysis biomass utilization	Did not analyze biomass utilization from integrated evaluation of economy, resource and environment
	Frank et al. *The Biomass assessment handbook*	2006	Summarized factors of biomass resource evaluation	Did not illustrate the comprehensive assessment of biomass resource utilization

utilization. But the research did not analyze biomass utilization from the view-points of the integrated evaluation of economy, resource and environment. And Frank et al. (2006) published the monograph of *The biomass assessment hand-book*, and summarized the factors of evaluation of biomass resource. However, the book just studied the quantity and quality of biomass resource, but did not illus-trate the comprehensive assessment of biomass resource utilization.

1.7.5 Utilization of Stockbreeding Wastes Treatment Technology

In China, the main technologies are Methane fermentation especially wet fermen-tation, composting and so on. In Japan, many advanced technologies have been adopted, for example, Methane fermentation, Sequence batch reactor, Tricking fil-ter bed, Drying method and so on, and received significant effects. In USA, the government adopted technologies of Oxidation pond and Constructed wetland to treat the stockbreeding waste. In European countries, they utilized the approaches of Drying method, Combustion method, Biochemical process etc. (see Table 1.3).

China Ministry of Agriculture (1999) draw up *The planning of energy and environment projects in large and medium-scaled livestock farms*, and calculated benefits of majority of domestic large and medium-scaled methane fermentation projects. It showed that the methane projects which utilized the biogas comprehen-sively have significant economic benefits. The investigation of this program calcu-lated and compared the economic benefit of different biomass waste technologies, but it did not analyze the resource and environment effects of stockbreeding waste utilization.

Catelo et al. (2008) researched the situation of family in the Philippines Free-range and large-scale farming pollution control, and adopted cost-benefit method to compare the methods of methane fermentation and organic fertilizer, according to net present value and sensitivity. The conclusion showed that methane fermen-tation is a more effective way. However, it did not analyze the synthetic benefits on the basis of resource and environment effects of stockbreeding waste utilization.

Huynh et al. (2007) studied the stockbreeding pollution of Vietnam and pointed out that the methane fermentation is the most effective technology from the view of both economic and environmental evaluation. The researcher indicated that the government should provide technical and financial supports to local residents, in order to promote methane fermentation. But this study did not assess the environ-ment from the comprehensive environmental pollution of stockbreeding waste, which contains atmospheric pollution, GHG and water pollution.

Studies about comprehensive environmental evaluation in China are mainly concerned with theories and theoretical model at national level. Researches on integrated practical assessment of environmental degradation and economic devel-opment are in initial stages. Spofford (1995) has made an attempt to estimate the impact of current environmental management policies on economic development in

Table 1.3 Utilization of stockbreeding waste treatment technology

Research	Author	Year	Achievements	Limitation
Stockbreeding biomass technology utilization	China ministry of agriculture. *The planning of energy and environment projects in large and medium-scaled livestock farms*	1999	The investigation of this program calculated and compared the economic benefit of different biomass waste technologies	Did not analyze the resource and environment effects of stockbreeding waste utilization
	Catelo et al.	2008	Researched the situation of family in the Philippines Free-range and large-scale farming pollution control, and adopted cost-benefit method to compare the methods of methane fermentation and organic fertilizer	Did not analyze the synthetic benefits based on resource and environment effects of stockbreeding waste utilization
	Huynh et al.	2007	Studied the stockbreeding pollution of Vietnam and pointed out that the methane fermentation is the most effective technology from view of economic and environmental evaluation	Did not assess the environment from comprehensive environmental evaluation of stockbreeding waste

China. Wang and Tang (2003) constructed a two-level theoretical model of environment and economic development to analyze the efficiency of environmental subsidies. The state environmental protection administration of China (2002) estimated the pollutants from stockbreeding wastes and the benefits of methane fermentation.

Moreover, present evaluation theories and methods of sustainable development are mainly concerned with sustainable development at global and national levels. However, the majority of the quantitative sustainable development evaluation emphasized particularly the construction of an indicator system and the comparative analysis of cities statically to evaluate the temporary state (Yan et al. 2010).

In most of these studies, conclusions are derived from simple data analysis and foreign experiences. There has been little research into the construction and analysis of a comprehensive simulation policy that is tailored to suit China's economy and social state and includes the introduction of present treatment technologies to control water pollutant emissions without deteriorating the socio-economic activities level, especially in the suburb around large cities. Therefore, feasible simulation should be constructed to realize the simultaneous pursuit of environmental preservation and economic development on the basis of characteristics of China.

In this article, we selected a typical suburb of big cities in China, Miyun County, Beijng. According to the basic simulation system and model constructed by Hirose and Higano (2000) and Mizunoya et al. (2006), we improved the simulation model and focused on the evaluation of water pollutant-minimization based on the ecological value of Miyun Reservoir. In the simulation, we considered the specific and special characteristics of China's economy and social state in terms of sustained economic growth rate, financial subordination relations and regional environmental policies which are different with the model of Japan. Besides these innovations above, we introduced two different advanced technologies from Japan and China to the study area through simulation with integrated policies and carried out regional analysis and allocation for two technologies which show strong operability in practice under the condition of limited funds and current states in China.

References

Arrow K (1995) Economic growth carrying capacity and the environment. Science 268:520–521
Ayres RU, Kneese AV (1969) Production, consumption and externalities. Am Econ Rev 59(3):282–297
Baumol WJ, Oates WE (1988) The theory of environmental policy. Cambridge University Press, UK, pp 156–190
Catelo MAO, Narrod CA, Tiongco MM (2008) Structural changes in the Philippine pig industry and their environmental implications. International Food Policy Research Institute (IFPRI), Washington, DC, p 781
China Ministry of Agriculture (1999) The planning of energy and environment projects in large and medium-scaled livestock farms. China Agriculture Press, China, pp 3–15
Coelho ST, Bolognini MF (1999) Policies to improve biomass-electricity generation in Brazil. Renew Energy 16:996–999
Forsund FR (1985) Input-output models: national economic models and the environment. In: Kneese AV, Sweeney JL (eds) Handbook of national resource and energy economics, vol 1, p 6

Frank RC, Sarah H, Peter DG (2006) The biomass assessment handbook: bioenergy for a sustainable environment. Earthscan Publications, London, pp 1–105

Gielen DJ, Fujino J, Hashimoto S, Moriguchi Y (2003) Modeling of global biomass policies. Biomass Bioenergy 25(2):177–195

Hadley G (1962) Linear programming. Addison-Wesley Press, New York, pp 10–54

Hannon B (2001) Ecological pricing and economic efficiency. Ecol Econ 36:19–30

Hektor B (2000) Planning models for bioenergy: some general observations and comments. Biomass Bioenergy 18:279–282

Higano Y, Sawada T (1997) The dynamic policy to improve the water quality of Lake Kasumigaura. Studies in Regional Science 26(1):75–86

Higano Y, Yoneta A (1999) Economical policies to relieve contamination of Lake Kasumigaura. Stud Reg Sci 29(3):205–218

Higano Y, Mizunoya T, Piao SH (2006) Ibaraki Preference in the city area project for promotion of coordination between industry, academia and government, and Advancement of Regional Science and Technology: the Lake Ksumigaura biomass recycling development project (2003–2005), pp 238–326

Hillier FS, Gerald JL (1967) Introduction to operations research. Holden-Day, Inc. Press, San Francisco, pp 107–121

Hillring B (1998) National strategies for stimulating the use of bioenergy: policy instruments in Sweden. Biomass Bioenergy 14(5):425–437

Hirose F, Higano Y (2000) A simulation analysis to reduce pollutants from the catchment area of Lake Kasumigaura. Stud Reg Sci 30(1):47–63

Huynh TTT, Aarnink AJA, Drucker A, Verstegen MWA (2007) Pig production in Cambodia, Laos, Philippines, and Vietnam: a review. Asian Journal of Agriculture and Development 4(1):69–90

Japanese Energy Association (2003) Biomass handbook, pp. 108–123

Jin D, Hoagland P, Dalton TM (2003) Linking economic and ecological models for a marine ecosystem. Ecol Econ 46:67–385

Jolliet O, Goedkoop M (2004) The LCIA midpoint-damage framework of the UNEP/SETAC life cycle initiative. Int J Life Cycle Assess 9(6):394–404

Lee KS (1982) A generalized input-output model of an economy with environmental protection. Rev Econ Stat 64(3):466–473

Leontief WW (1977) The future of the world economy. Oxford University Press, New York, pp 103–124

Mandal KG, Saha KP, Ghosh PK, Hati KM, Bandyopadhyay KK (2002) Bioenergy and economic analysis of soybean-based crop production systems in central India. Biomass Bioenergy 23:337–345

Mitchell CP (2000) Development of decision support systems for bioenergy applications. Biomass Bioenergy 18:265–278

Mizunoya T, Sakurai K, Kobayashi S, Piao SH, Higano Y (2006) A simulation analysis of synthetic environment policy: effective utilization of biomass resources and reduction of environmental burdens in Kasumigaura basin. Stud Reg Sci 36(2):355–374

Moffatt I, Hanley N (2001) Modeling sustainable development: systems dynamic and input-output approaches. Environ Model Softw 16:545–557

Oka T (1997) Welfare economics and environmental policy. Iwanami Press, Tokyo, pp 91–102

Perrings C (1987) Economy and environment. Cambridge University Press, New York, pp 110–165

Ramachandra T, Joshi N, Subramanian D (2000) Present and prospective role of bioenergy in regional energy system. Renew Sustain Energy Rev 4(4):375–430

Ray DE (2000) Biomass and bioenergy applications of the POLYSIS modeling framework. Biomass Bioenergy 18:291–308

Seungdo K, Bruce ED (1999) Cumulative energy and global warming impact from the production of biomass for biobased products. J Ind Ecol 7(3):147–162

Spofford WO (1995) Integrating environmental management and economic development in China. Resources 119:102–109

State Environmental Protection Administration (2002) The pollution and measurements for stockbreeding industry wastes in China. China Environmental Science Press, Beijing, pp 12–31, 40–55

State Environmental Protection Administration (2008) State of the environment report. China Environmental Science Press, Beijing, pp 10–32, 51–74

Takagi A (1999) Economic evaluation of the water quality improvement policy in the closed water area. Stud Environ Syst 27:9–16

Turner RK (2000) Integrating natural and socio-economic science in coastal management. J Mar Syst 25:447–460

Wagner HM (1959) Linear programming techniques for regression analysis. J Am Stat Assoc 57:206–212

Walsh ME, Ugarte DG, Shapouri H, Slinsky SP (2003) Bioenergy crop production in the United States: potential quantities, land use changes, and economic impacts on the agricultural sector. Environ Resource Econ 24(4):313–333

Wang B, Tang LJ (2003) Analysis of environmental and economic integration systems optimization. Systems Engineering-theory Practice 3:14–17

Yan JJ, Kang CJ, Xu F, Higano Y (2010) The synthetic quantitative evaluation of sustainable development of resource-based cities: a case study of Panjin City. Studies in Regional Science 40(2):353–372

Chapter 2
Characteristics of the Study Area

Abstract Miyun County is the largest rural suburb and the most important county of Beijing. Miyun County has a unique significance for the regional environment, and dramatically influences the water safety and local development of Beijing. In the catchment areas of Miyun Reservoir, stockbreeding wastes are generated and heavily contribute to water pollution, especially pig farming, which has become the source of the most serious pollutants of Miyun Reservoir. In this Chapter, the characteristics of the Miyun County, Miyun Reservoir, and the study area, and the current environmental conditions are presented.

Keywords Miyun County · Miyun Reservoir · Characteristics

2.1 Development of Stockbreeding Industry and Its Pollution in Beijing

2.1.1 Natural Geographical Environment

Beijing is the capital of China and the political and cultural centre of China. it is located at latitude between 39°26′ and 41°03′N and longitude between 115°25′ and 117°30′E. It lies in the north of China and covers an area of 16.8 thousand square kilometers. January, with an average temperature of −4.7 °C, is the coldest month and July, with an average temperature of 26.1 °C, is the hottest month. The rural areas of Beijing consists of 10 rural areas about 15,800 and 10,930 km^2 is agricultural land.

The total population of Beijing is 11.2 million in 2007, and the density of population is 668 people per km^2, which is more than 5 times of the average level of China. The population of rural resident is 8.83 million, which accounting for 78.8 % of the whole city population and leading to the huge pressure.

Yan J.J., Xu F., Higano Y., 2010c. Comprehensive evaluation of integrated pollutant-minimization policies in rural area around Beijing: case study of Miyun County. Journal of Human and Environmental Symbiosis, no.17, pp 87–98.

© The Author(s) 2015
J. Yan, *Comprehensive Evaluation of Effective Biomass Resource Utilization and Optimal Environmental Policies*, SpringerBriefs in Economics, DOI 10.1007/978-3-662-44454-2_2

Beijing Natural channel consists of five water systems from west to east: Jumahe River water system, Yongdinghe River water system, Beiyunhe water system, Chaobai River water system, Jiyunhe water system. There are no natural lakes in Beijing. There are 85 reservoirs, of which Miyun Reservoir and Guanting Reservoir, Huairou Reservoir, Haizi Reservoir are large reservoirs. Miyun Reservoir is the only surface water resource of Beijing.

2.1.2 Development of Stockbreeding Industry in Beijing

The gross output value of stockbreeding industry was 12.36 billion RMB Yuan in 2005, which accounted for 45.78 % of the total agricultural output value (Fig. 2.1). And the gross income of stockbreeding industry was 9.57 billion RMB in 2005, which accounted for 45 % of total agricultural income (Fig. 2.2). The population

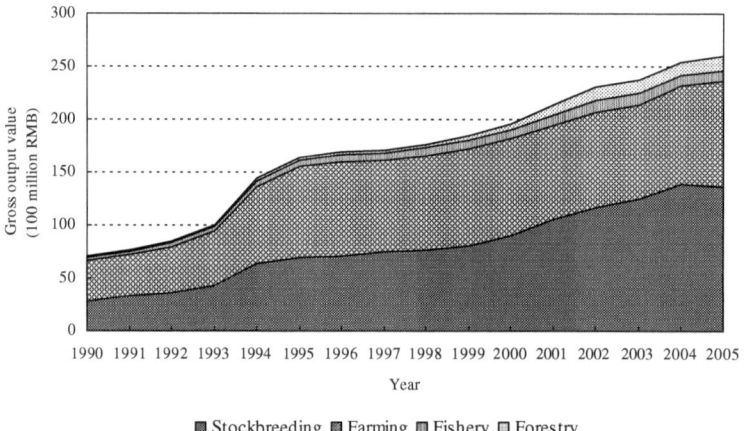

Fig. 2.1 The gross output value of farming, forestry, stockbreeding and fishery industry (1990–2005). *Source* Beijing Statistics (2006), pp 23–24

Fig. 2.2 The distribution of the first industry production in rural areas of Beijing (2005) *Source* Beijing Statistics (2006), pp 8–10

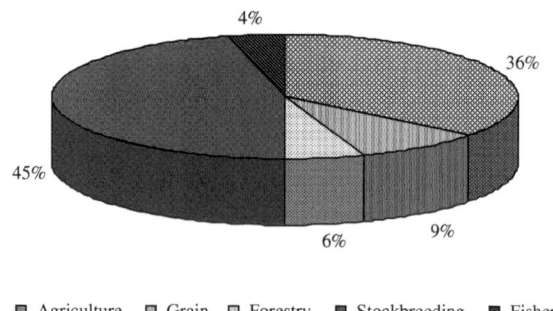

Table 2.1 The number of stockbreeding animals (2005)

Number of stockbreeding	Number
Beef cattle slaughtered in the whole year	24.5
Pork pigs slaughtered in the whole year	448.7
Sheep slaughtered in the whole year	256.1
Livestock on hand (year-end)	3,125.1

Source Beijing Statistics (2006), pp 18–19
Unit: 10,000 heads

Table 2.2 The output of stockbreeding products (2005)

Output of stockbreeding products	Output
Output of meat	66.7
Pork	31.3
Beef	4.4
Mutton	4
Cow milk	64.2
Poultry eggs	16
Eggs	15.6
Honey	3,282

Source Beijing Statistics (2006), pp 16–17
Unit: 10,000 t

engaged in stockbreeding industry was 77,042 people in 2005, which solving the employment problem nearly 100,000 people.

Table 2.1 represents the number of stockbreeding animals in 2005. The number of Large Animals on hand (year-end) was 266,000 heads; Pork Pigs Slaughtered in the whole year was 4,487,000 heads; Beef Cattle Slaughtered in the whole year was 245,000 heads; Sheep Slaughtered in the whole year was 2,561,000 heads; Livestock on Hand (year-end) was 31,251,000 heads in 2005.

Table 2.2 indicates the gross output of stockbreeding waste was 3 million tons in 2005. There are two treatment methods of stockbreeding waste: (1) majority of the wastes are treated by dry haul treatment method; (2) the method of water jetting concentrated in large-scale breeding farm. The general utilization method of stockbreeding waste is returning to the stockbreeding waste to the field, and the adoption of infiltration to treat the sewage discharge.

2.1.3 Environmental Pollution Caused by Stockbreeding Wastes

According to sewage capacity coefficient that calculated by the State Environmental Protection Administration (2002), the gross output of stockbreeding waste was 3,044,200 t in 2005, the total net load of COD was 93,400 t, T-N was 18,500 t and T-P was 7,000 t, respectively (Table 2.3).

Table 2.3 Water pollutants emitted by stockbreeding wastes (2005)

Water pollutants	Pollutant
COD	93,400
T-N	18,500
T-P	7,000

Source Beijing Statistics 2006, pp 57–59
Unit: t

Table 2.4 Air pollution emitted by stockbreeding wastes (2005)

Air pollutants	Pollutant
N_2O	1,125.3
CH_4	64,209.54
NH_3	38,647.87

Source Beijing Statistics 2006, pp 57–59
Unit: t

According to IPCC (2006) standards, the emission of air pollution of N_2O that caused by stockbreeding waste was 1,125.3 t/year, NH_3 was 38,647.87 t/year, CH_4 was 64,209.54 t/year, respectively (Table 2.4).

The average loading quantity of stockbreeding waste in the cultivated land of the rural areas of Beijing has reached 30.9 t per 10,000 m^2. Comparing with the situation in the suburbs of Shanghai and Zhejiang Province, the inequality extent of Beijing is more serious than other big cities.

2.1.4 Potential Value of Biomass Resource of Stockbreeding Wastes

There is massive utilizable nutrition in stockbreeding wastes biomass, including nutrients of N, P and K. And all the utilizable nutrition can be utilized as biomass resource that is very effective and useable to land and farming by proper treatment methods.

2.2 Miyun County and Miyun Reservoir

2.2.1 The Importance of Miyun County and Miyun Reservoir

Miyun County is the largest suburb and the most important county of Beijing with an area of 2,229 km^2 and about 426 thousand people. The gross regional product (GRP) was about 5,126 million RMB Yuan in 2006. Especially Miyun Reservoir, the only surface water resource of Beijing, provides more than 60 % of the water resources for Beijing. Miyun Reservoir, located 100 km northeast of Beijing City,

Table 2.5 Physical dimensions of Miyun reservoir

Surface area	188 km^2
Volume	0.4375 km^3
Maximum depth	43.5 m
Catchment area	15,788 km^2

Source Miyun County Statistics (2007), pp 14–15

is a mountain valley reservoir. It was built in September of 1960, and is the largest reservoir in Beijing area. Two main rivers, Chao River and Bai River, flow into Miyun Reservoir. The total catchment is about 15,788 km^2, consists of mountains and piedmonts and lacks large industrial enterprises. Sediments and nutrients from agricultural, pastoral and forestry lands that enter the reservoir with overland flow make the main pollution sources (Table 2.5).

The initial purpose of building Miyun Reservoir was flood control, irrigation and fishery, but it has become more and more important as main drinking water storage for Beijing area. The water quality is therefore of great importance. Recently, the nutrient concentration and the number of phytoplankton in the Reservoir are constantly rising to arrest the attention of Beijing municipality. A project for keeping the water clean is being put into effect.

This important ecological value of Miyun County has unique significance for the region's environment and dramatically influences water quality and local development in Beijing.

2.2.2 The State of Water Environment in the Catchment Area

Though T-N (total nitrogen), T-P (total phosphorus) and COD (chemical oxygen demand) contents of the lake water are low, the population densities of plankton and benthic invertebrates have increased as compared with those in 1981, indicating a trend toward eutrophication. The concentration of COD and T-N gradually increased during the year from 1994 to 2006.

Stockbreeding wastes in the catchment areas of Miyun Reservoir contribute heavily to water pollution. Figure 2.3 shows the detailed structure of water pollutants of the catchment area of Miyun Reservoir, the amount of T-N emitted by pig farming industry accounted about 25 % of total amount of T-N. About 41 % of T-P and 36 % of COD were emitted by pig farming industry (Figs. 2.4, 2.5). Therefore, pig farming industry has become the source of the most serious pollutants to Miyun Reservoir.

Future development of Miyun County and decreases in annual runoff to Miyun Reservoir are expected to significantly influence the regional environment and the water quality of Miyun Reservoir worsen, leading to a serious water safety crisis for Beijing.

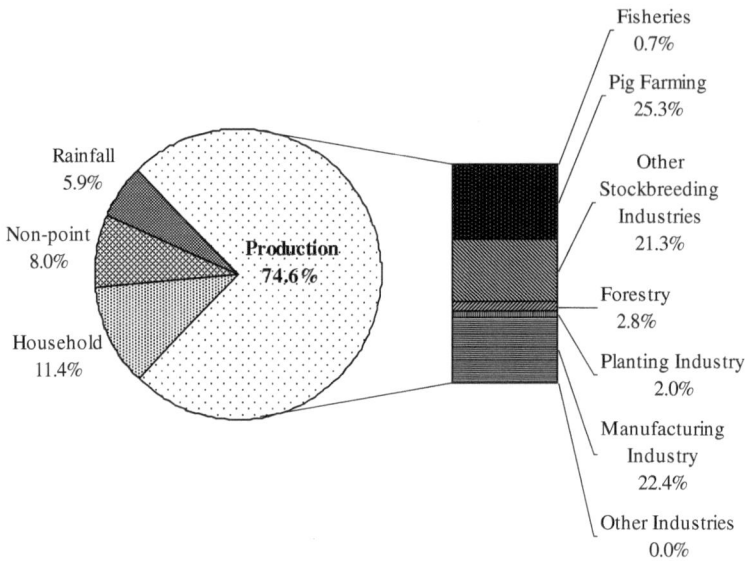

Fig. 2.3 Proportion of the net load of T-N in the catchment area. *Source* Miyun County Statistics (2007), pp 16–17

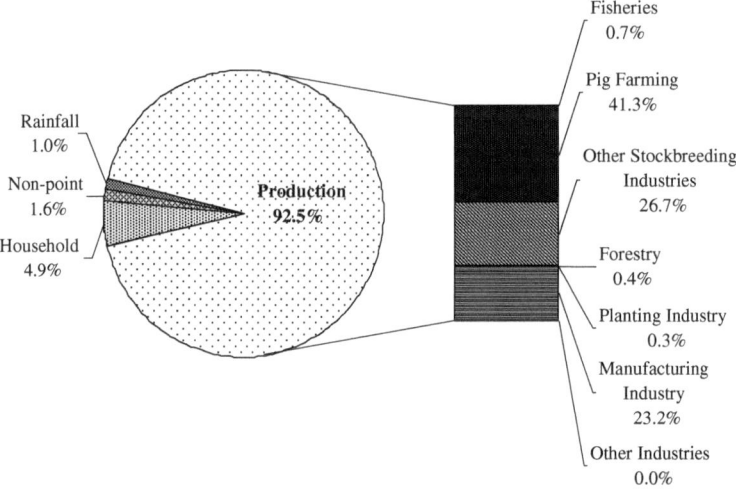

Fig. 2.4 Proportion of the net load of T-P in the catchment area. *Source* Miyun County Statistics (2007), pp 16–17

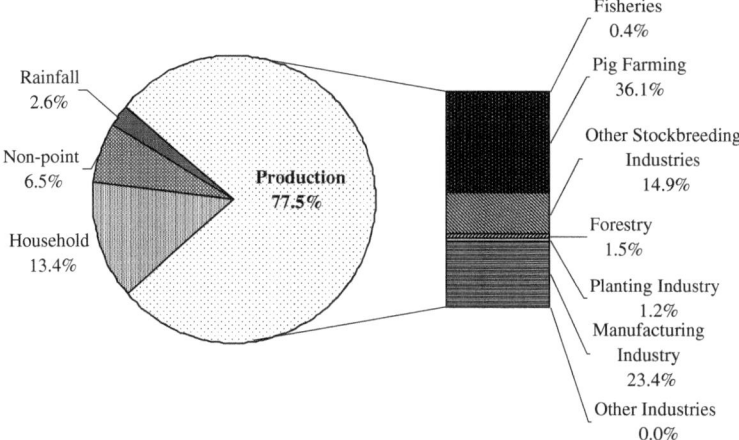

Fig. 2.5 Proportion of the net load of COD in the catchment area. *Source* Miyun County Statistics (2007), pp 16–17

2.2.3 The State of Water Environmental Management in the Catchment Area

Since Miyun Reservoir is one of the most important water sources for Beijing, fishery may be allowed with certain limitation as long as it causes no harm to water quality. Water recreations are all forbidden. In the catchment area, water and soil conservation practices are encouraged.

The catchment area introduced wet methane fermentation technology and composting to treat serious water pollutants emitted by the stockbreeding industry. The government also proposed and provided preferential prices for farmers to promote utilization of organic fertilizer. In order to conserve water and soil for the Miyun Reservoir, the local government also considered the policies of a water conservation forest to conserve water resources for the reservoir and minimize pollutants indirectly based on the requirements of the region.

On the other hand, in order to satisfy the demand for animal products of Beijing and improve the local economic development, the policy of "Strong Stockbreeding County" has been raised by Miyun government. It is can be expected that the stockbreeding industry would dramatically increase in the future 5–10 years which would result in the serious environmental pollution without proper treatment methods for stockbreeding wastes. There is no reason to think that we must deteriorate the socio-economic activity level to improve the quality of environment.

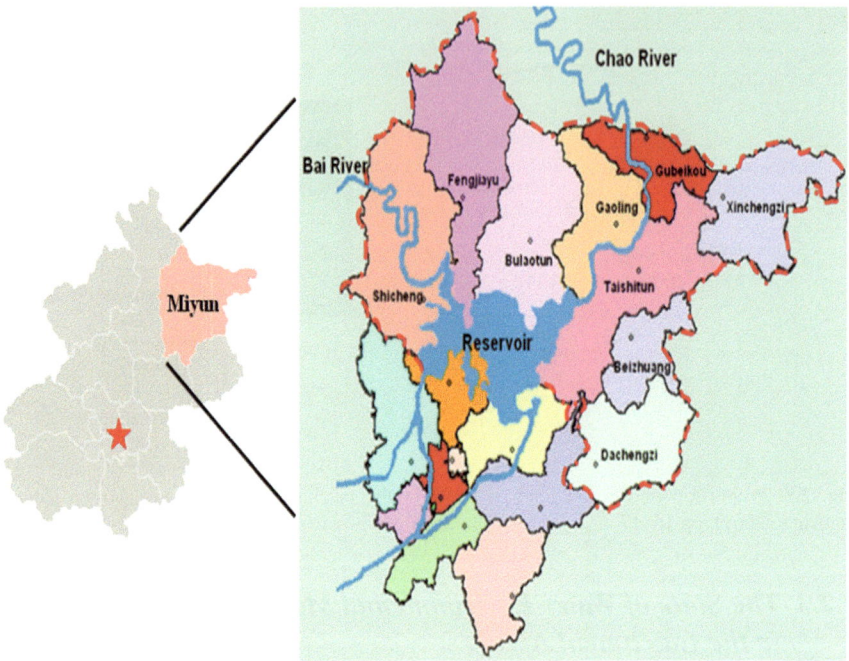

Fig. 2.6 Location of the study area

2.3 Study Area: The Upstream Area of Miyun Reservoir in Miyun County

The study area covers the upstream area of Miyun Reservoir in Miyun County. In this study area, there are nine towns of Miyun County. Two main rivers, Chao River and Bai River flow through the nine towns into Miyun Reservoir (Fig. 2.6). There area 12.9 thousand people in the study area. The gross regional product (GRP) was about 1,948 million RMB Yuan in 2006 which accounted about 38 % of the total GRP of Miyun County.

2.3.1 Classification of Sub-basins of the Main Rivers and Water Pollutant Sources

The Chao and Bai rivers are the main rivers flowing into the Miyun Reservoir. The two rivers flow through nine towns of Miyun County. These rivers are structured into the model, and other small rivers in the sub-basins flow into one of these two major rivers (Table 2.6). We measured three indicators of water pollution, including T-N (total nitrogen), T-P (total phosphorous) and COD (chemical oxygen demand) (Table 2.8) (Yan et al. 2010b).

Table 2.6 Classification of the study area

Region index	Rivers		Towns	
	Index	Name	Index	Name
1	1	Bai river	1	Shi cheng
			2	Feng jiayu
			3	Bu laotun
2	2	Chao river	4	Gao ling
			5	Gu beikou
			6	Xin chengzi
			7	Tai shitun
			8	Bei zhuang
			9	Da chengzi

Table 2.7 Population of the study area in 2005 and 2006

Basin	Index	Town name	2006	2005	Variation
Bai river	1	Shi cheng	5,522	5,647	−125
	2	Feng jiayu	9,266	9,799	−533
	3	Bu laotun	23.478	23,937	−459
		Total of Bai river basin	38,266	39,383	−1,117
Chao river	4	Gao ling	17,837	17,736	101
	5	Gu beikou	8,226	8,478	−252
	6	Xin chengzi	12,223	12,341	−118
	7	Tai shitun	26,598	27,024	−426
	8	Bei zhuang	8,543	8,443	100
	9	Da chengzi	16,959	17,168	−209
		Total of Chao river basin	90,386	91,190	−804
		Total of the catchment area	128,652	130,573	−1,921

Source Miyun County Statistics (2007), pp 10–11
Unit: person

2.3.2 Regional Socio-economic Status of the Study Area

Although the upstream area of Miyun Reservoir is the most important area of Miyun County, the gross regional product (GRP) of the study area was about 1,948 million RMB Yuan in 2006 which accounted about 38 % of the total GRP of Miyun County. The total population of the study area was about 12.9 thousand in 2006 and it decreased about 1.47 % as compared to the population in 2005 (Table 2.7).

In this area, Chao River Basin is the larger basin occupying 50 % of the total area in the whole basin. Approximately 90,000 people live in the basin indicating about 70 % of total population of the study area. About 79.5 % of the total output value was produced by Chao River Basin.

Industry in study area is classified according to fisheries, pig farming industry, other stockbreeding industry, forestry, planting industry, manufacturing industry and other industries as shown in Table 2.8. From this table we can conclude that

Table 2.8 Output value of industries in the study area of 2006

Basin	Index	Town name	Total output value	Fisheries	Pig farming industry	Other stock-breeding industries	Forestry	Planting industry	Manufacturing industry	Other industries
Bai river	1	Shi cheng	14,526	399	212	2,330	832	135	6,280	4,337
	2	Feng jiayu	28,162	70	780	5,058	792	747	17,372	3,343
	3	Bu laotun	45,022	1,238	3,726	6,937	798	1,411	23,160	7,753
		Total of Bai river basin	87,710	1,707	4,719	14,325	2,422	2,293	46,812	15,433
Chao river	4	Gao ling	43,223	240	4,080	5,993	907	1,420	23,610	6,974
	5	Gu beikou	22,518	100	1,194	2,670	598	289	10,515	7,153
	6	Xin chengzi	41,593	160	1,370	2,883	581	1,164	23,472	11,964
	7	Tai shitun	120,419	1,306	2,813	7,695	1,799	2,272	77,303	27,231
	8	Bei zhuang	68,071	319	1,591	9,321	603	477	52,485	3,274
	9	Da chengzi	43,874	399	1,979	4,511	653	985	31,100	4,246
		Total of Chao river basin	339,698	2,523	13,027	33,072	5,142	6,607	218,485	60,842
		Total of the catchment area	427,408	4,230	17,746	47,396	7,564	8,900	265,297	76,275

Source Miyun County Statistics (2007), pp 12–13

Unit: 10,000 RMB Yuan

Table 2.9 The area of different land use in 2006

Basin	Index	Town name	Upland cropping	Forest land	Orchard land	City area	Other land area	Total area
Bai river	1	Shi cheng	0.75	17.75	1.51	34.65	8.35	63.01
	2	Feng jiayu	2.38	17.48	9.13	25.06	8.60	62.65
	3	Bu laotun	4.72	16.14	25.31	1.49	2.03	49.69
	Total of Bai river basin		7.85	51.37	35.95	61.20	18.97	175.35
Chao river	4	Gao ling	7.16	11.23	20.41	1.24	1.39	41.42
	5	Gu beikou	2.47	12.71	2.13	7.23	4.37	28.91
	6	Xin chengzi	6.03	8.86	14.54	6.58	7.83	43.83
	7	Tai shitun	7.37	24.30	27.97	1.88	1.07	62.59
	8	Bei zhuang	3.07	9.12	4.54	5.39	4.85	26.97
	9	Da chengzi	2.92	9.43	11.02	12.92	6.77	43.05
	Total of Chao river basin		29.01	75.65	80.60	35.23	26.28	246.77
Total of the catchment area			36.86	127.02	116.55	96.44	45.25	422.12

Source Miyun County Statistics (2007), pp 8–10
Unit: km^2

manufacturing industry is the most important economic activity in the study area, which contributed about 62 % of the total output value in 2006. However, the local government has already raised the policy to reduce and limit the scale of manu-facturing industry in order to protect the environment of upstream area of Miyun Reservoir. Pig farming industry and other stockbreeding industry are also impor-tant economic activities in the study area, which accounted about 4 and 11 % of the total output value of the area respectively. With the demand for animal prod-ucts and a policy of "Strong Stockbreeding County" by Miyun government, the production of stockbreeding industry would dramatically increase in the future.

Land use in the study area is classified into five categories, including upland crop-ping, forest land, orchard, city area and other land area as shown in Table 2.9. It is concluded from the table that, distribution of the land use pattern is different in terms of upland cropping, orchard and city area in basins. City area are mainly be occupied by Bai River Basin. In addition, Miyun government has already raised the policy of Organic Fertilizer Promotion Policy in 2004 and converted majority of upland crop-ping into forest land to improve water conservation function of Miyun reservoir, the area of forest land accounted about 30 % of the total area in the study area in 2006.

2.3.3 Total Pollution in the Study Area

The total amounts of pollutants flowing into the study area were estimated using official data provided by the state environmental protection administration of China (2002) and Yearbook of Miyun County (2007).

Table 2.10 Net load of water pollutants by different sources in the study area (2006)

Sources of pollutants	Frame	Unit	Net load of pollutants (t/year)		
			T-N	T-P	COD
1. Household					
Sewage plant	10,309	Person	10.95	1.64	54.75
Untreated waste water	118,343	Person	395.95	40.22	2,002.99
Total	128,652	Person	406.90	41.87	2,057.74
2. Non-point					
Upland cropping	36.864	km^2	30.59	1.20	55.30
Forest land	127.018	km^2	72.32	2.50	177.57
Orchard land	116.545	km^2	69.68	2.41	171.08
City area	96.437	km^2	84.48	6.34	538.55
Other land area	45.254	km^2	29.40	1.15	62.36
Total	422.119	km^2	286.48	13.60	1,004.85
3. Production					
Fisheries	4,230	10^4 RMB Yuan	25.56	5.65	64.06
Pig farming	17,746	10^4 RMB Yuan	901.18	355.15	5,544.80
Other stockbreeding industries	47,396	10^4 RMB Yuan	758.04	229.77	2,291.31
Forestry	8,900	10^4 RMB Yuan	100.27	3.61	226.37
Planting industry	7,564	10^4 RMB Yuan	72.32	2.50	177.57
Manufacturing industry	265,297	10^4 RMB Yuan	797.52	199.38	3,588.86
Other industries	76,275	10^4 RMB Yuan	0	0	0
Total	427,408	10^4 RMB Yuan	2,654.90	796.06	11,892.54
4. Rainfall			211.35	8.92	393.19
Total	188	km^2	211.35	8.92	393.19
Total			3,559.62	860.45	15,348.33

Source Miyun County Statistics (2007), pp 10–25

We used the data of 2006 for the study area. The basic data were obtained from the Yearbook of Miyun County (2007) and the digital data for the simulations were calculated based on this basic data and related statistics from the State Environmental Protection Administration (2002).

From the Table 2.10, we can conclude that the net load of waster pollutants by production occupied about 75 % of T-N, 92 % of T-P and 77 % of COD respectively. Therefore, the production has become the main water pollution source of the study area.

According to the results of statistics, stockbreeding wastes in the study area contribute heavily to water pollution, especially pig farming has become the source of the most serious pollution in the Miyun Reservoir. The net load of T-N, T-P and COD of pig farming industry accounted about 34, 45 and 47 % of the total net load of production respectively (Figs. 2.7, 2.8 and 2.9). In addition, with the implementation of "Strong Stockbreeding County" policy, it is can be expected that the pollutants emitted by pig farming wastes would dramatically increase in the future. Therefore, we should raise and carry out proper treatment method and

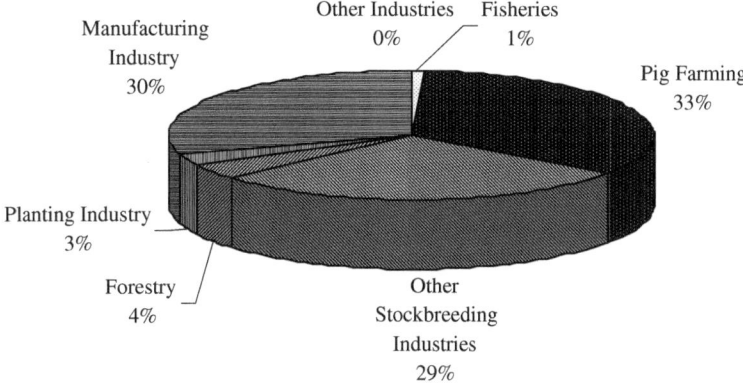

Fig. 2.7 Proportion of the net load of T-N in industry in the study area. *Source* Miyun County Statistics (2007), pp 19–20

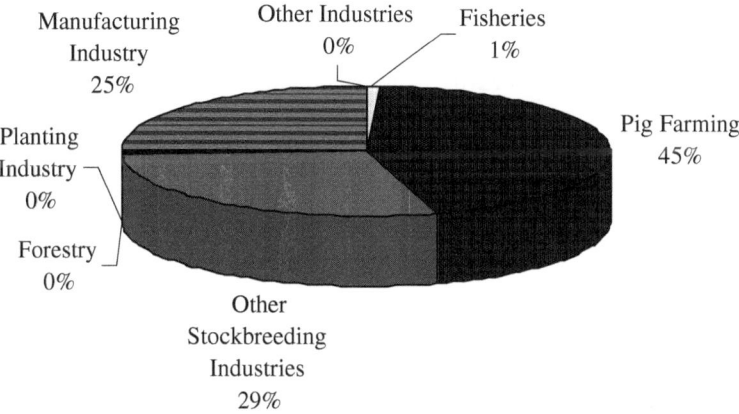

Fig. 2.8 Proportion of the net load of T-P in industry in the study area. *Source* Miyun County Statistics (2007), pp 19–20

integrated policies to realize the simultaneous pursuit of environmental preservation and economic development in the study area.

2.4 Implications and Conclusions

In this chapter, the characteristics of the Miyun County, Miyun Reservoir and the study area and the current environmental conditions are presented. In particular, this caused of water pollution are detailed. The main points conducted are: (1) stockbreeding industry specially pig farming industry concentrate in the catchment area of Miyun Reservoir causing considerable decline of water quality;

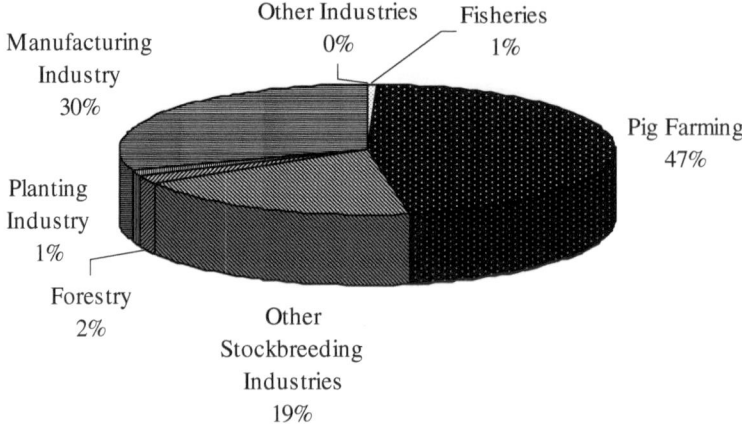

Fig. 2.9 Proportion of the net load of COD in industry in the study area. *Source* Miyun County Statistics (2007), pp 19–20

(2) stockbreeding wastes in the study area (upstream area of Miyun Reservoir in Miyun County) contribute heavily to water pollution, especially pig farming has become the source of the most serious pollution in the study area; (3) because of the requirements of water environmental protection in Miyun Reservoir, the regional economy slowly developed in the recent years in the study area; (4) Studies alert that, decreases in annual runoff to Miyun Reservoir are expected to significantly influence the regional environment, and the water quality of Miyun Reservoir worsen, leading to a serious water safety crisis for Beijing.

References

Beijing Statistics (2006) Beijing statistical yearbook. China's Statistics Press, Beijing, pp 8–10, 16–19, 23–24, 39–40, 57–60
IPCC (2006) IPCC guidelines for national greenhouse gas inventories. http://www.ipcc-nggip.iges.or.jp/public/2006gl/index.html, pp 212
Miyun County Statistics (2007) Yearbook of Miyun County statistics. China's Statistics Press, Beijing, pp 5–20, 25–30, 46–50
State Environmental Protection Administration (2002) The pollution and measurements for stockbreeding industry wastes in China. China Environmental Science Press, Beijing, pp 12–31, 40–55
Yan JJ, Xu F, Kang CJ, Higano Y (2010) Effective stockbreeding biomass resource use and its impact on water environment from the viewpoint of sustainable development. J Dev Sustain Agri 5:147–150

Chapter 3
Evaluation of Water Pollutant Minimization Policies with Adopting Present Technology

Abstract In this Chapter, we adopted the integrated policies to minimize water pollutants flowing into Miyun Reservoir based on two sub-models. Particularly, we improved the economic models so that they are more adaptive to the specific characteristics and the actual situation of suburbs of big cities in China. Based on the results of the simulation, the efficiency of wet methane fermentation technology is not sufficient to treat the high concentrations of wastes in the water and meet the requirements of future economic development in the catchment area. Especially, the pig farming industry requires the introduction of advanced technology to allow simultaneous pursuit of environmental preservation and economic development.

Keywords Integrated policies · Water pollutants · Present technology

On the basis of data collected through a sample survey, we have gathered relevant data on the socio-economic and water pollution situation in the area. Moreover, studying other researches, we considered both the ecological system in the objective area and the socio-economic situational changes during a specific period that have general applicability to the rural areas of China. To achieve our research purposes, a water pollutants flow balance model, a socio-economic model and one objective function (to minimize total nitrogen (T-N) levels over 10 years) were specified to express all the key factors and parameters reflecting the environmental situation and affecting human activities. We formulate the comprehensive system as a linear optimization model which we solve a mathematical optimization software package, LINGO. This enables us to simultaneously simulate the water environmental system and the socio-economic system by creating as much linearity as possible in the functions that is close to the reality. Detailed discussions of linear programming theory may be found in books by Hadley (1962), Hillier and Gerald (1967) and Wagner (1959).

Yan J.J., Xu F., Higano Y., 2010c. Comprehensive evaluation of integrated pollutant-minimization policies in rural area around Beijing: case study of Miyun County. Journal of Human and Environmental Symbiosis, no. 17, pp. 87–98.

© The Author(s) 2015
J. Yan, *Comprehensive Evaluation of Effective Biomass Resource Utilization and Optimal Environmental Policies*, SpringerBriefs in Economics,
DOI 10.1007/978-3-662-44454-2_3

3.1 Model Characteristics and Structure

3.1.1 Framework of the Model

The dynamic simulation model consisted of a water pollutants flow balance model, socio-economic model and one objective function. The policy was derived to minimize T-N, which was subjected to sub-models. The water pollutants flow balance model describes how the pollutants flow into the rivers and reservoir. The socio-economic model represents the social and economic activities in the region and the relationship between the activities and emission of pollutants. The composition of the sub-models is shown in Fig. 3.1.

3.1.2 The Classification of Each Generation Source, Basic Data and Simulation Running Period

The pollutants measured in this study area were three water pollutants total nitrogen (T-N), total phosphorus (T-P) and chemical oxygen demand (COD). We classified the water pollutant generation sources as household, non-point, production and pig farming. Household wastewater disposal systems were classified into two categories based on the situations in the suburbs; land use was classified into five categories of non-point generation sources and industry was classified into seven

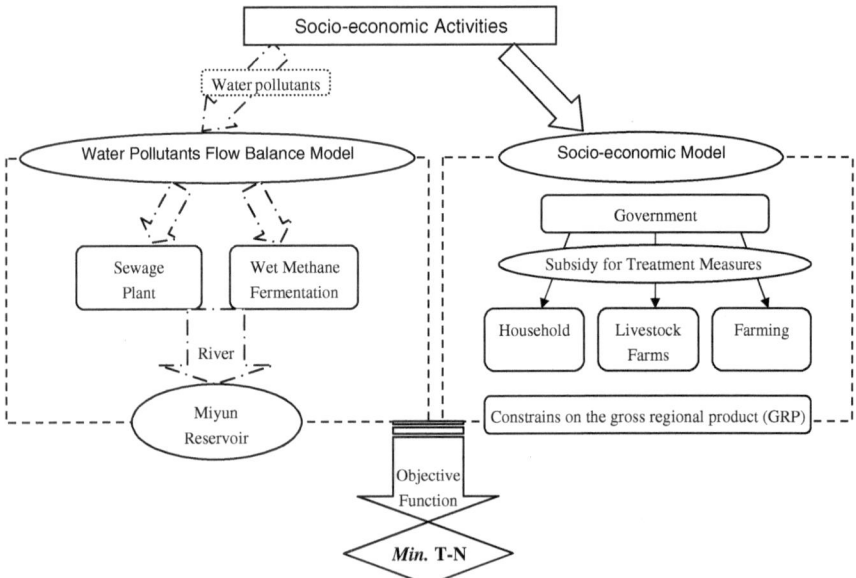

Fig. 3.1 Framework of the pollution-minimization model

Table 3.1 Classifications of household waste water disposal system

Index	Facility
1	Sewage plant
2	Untreated waste water

Table 3.2 Classification of industry

Index	Industry
1	Fisheries
2	Pig farming
3	Other stockbreeding industries
4	Forestry
5	Planting industry
6	Manufacturing industry
7	Other industries

Table 3.3 Classification of land use

Index	Land use
1	Upland cropping
2	Forest land
3	Orchard land
4	City area
5	Other land area

categories of production generation based sources on the characteristics of Miyun County (Tables 3.1, 3.2 and 3.3). "Other land use" and "Other industry" showed no water pollutant emission. We used the data of 2006 and a simulation running period from 2007 to 2016 (total 10 years).

3.1.3 The Hypotheses of Simulation Model

3.1.3.1 Sub-basin of the Major Rivers and Municipality in the Study Area

The basin is classified into two sub-basins of two main rivers (Chao River and Bai River). When the area of a municipality belongs to both of sub-basins, the ratio of the area that belongs to the sub-basin is calculated and it is assumed the ratio of rainfall in the sub-basin to the total in the municipality is same as that ratio, and all the rainfall in the sub-basin flows into the main river of the sub-basin. Also, we assume that small rivers in the sub-basin all flows into the main river of the sub-basin.

3.1.3.2 Policy Measures Against Water Pollution

Policies against water pollution listed in Table 3.4 are currently implemented by the Miyun Government. Those are installed in the simulation model as control variables and it is assumed that the Government must select the best combination of policies and the best assignment of budget for implementation of the best policies in the dynamic and spatial context in order to improve water quality in the Reservoir as much as possible while keeping the economic growth in the basin as much as possible. It is also assumed that the annual budget with which the government can implement the best policies is limited to 149.3 million RMB Yuan.

3.1.3.3 Skeleton of the Simulation Model

The model is composed of two sub-models and one objective function. The objective function is minimization of the total nitrogen that flows into the reservoir. The socio-economic sub-model describes socio-economic activities in the basin and the water pollutants are linearly dependent on the level of the socio-economic activities in the sub-basins. In order to simulate dynamic motivation in the market with the sub-model, it is added to the sub-model as a constraint that the GRP cannot decrease every year, which is a plausible assumption that considering the

Table 3.4 Integrated policies for the study area

Objectives	Name of policy	Index	Measurements
Production	Reduction of the capital employed	1	Subsidization for industries to reduce working capital and thus adjust production
Stockbreeding industry (pig farming)	Reduction of the capital employed and adoption of wet methane fermentation technology	2	Subsidization for stockbreeding industry to reduce capital stock so as to adjust production
		3	Adoption of general wet methane fermentation technology
Household	Sewage system and sewage plant	4	Subsidization for the municipality to install more sewage systems and sewage plants
Non-point	Promotion of organic fertilizer	5	Subsidization for farmers who uses organic fertilizer
Water conservation	Water conservation forest	6	Subsidy to promote water conservation forest (conversion of other land to forest)

Source Yan et al. (2010, pp. 1–12)

current situation in China. The sub-model of socio-economic activities is an open model with import and export of commodities. It is assumed that the import and export of commodities are limited within the intervals that are calculated based on the actual data. However, stock breeding wastes are not imported nor exported. Those produced in the basin must be treated in it.

3.1.4 Water Pollution-Minimization Policies

Table 3.4 presents the integrated policies used in the simulation model to reduce water pollutants and conserve water resource for the Miyun Reservoir, and the government selected policies used to efficiently improve the regional water environment. Particularly, the catchment area introduced wet methane fermentation technology to treat serious water pollutants emitted by the stockbreeding industry. The government also proposed and provided preferential prices for farmers to promote utilization of organic fertilizer. We also considered the policies for a water conservation forest to conserve water resources for the reservoir and minimize pollutants indirectly based on the requirements of the region.

3.2 Principles of Modeling and Important Indicators

3.2.1 Principles of Modeling

The relationship between pollution and economic activity is described by a simple material balance model. In an economy with no imports and exports and where there is no net accumulation of stocks (plant, equipment, inventories, consumer durables or residential buildings) the mass of residuals returned to natural environment must be equal to the mass of basic fuels, food minerals and other raw materials entering the processing and production system plus gases taken from the atmosphere. This is the principal of material balance (Freeman et al. 1973). In this study, the basic modeling principle is the processing and production activities described in economic terms such as market flow using input–output (I–O) table with macroeconomic indicators and the relation between production, consumption, investment, capital stock, value added and the environmental media that receive residuals as wastewater after whole processing. The environment is a provider of materials and services to the economy. We shall see that this return flow has an adverse impact on the volume and quality of other environmental services. We focus on the flow of material from the environment to the economy and the return flow of these materials back to the environment as wastes or residuals. In particular, the principal of the modeling is to minimize the volume of water pollutants return to the environment.

3.2.2 *Important Indicators*

The indicators are applied that use commonly in the world-wide in the literature related to water pollution such as chemical oxygen demand (COD), total nitrogen (T-N) and total phosphorus (T-P) known as organic pollution parameters.

COD is an index which measure how clear or polluted due to organic materials. The waste water is mostly includes material that can be grasped in COD. It is a measure of water quality using as an index of lake, wetlands, and seawater pollution by organic matter. Chronologically, firstly index COD adopted after that T-N and T-P had been adopted. When the standard is set in only COD, the cost of the firm to process the waste water in order to clear the effluent standard is lower than when standard in T-N and T-P are added.

In environmental chemistry, the chemical oxygen demand (COD) test is commonly used to indirectly measure the amount of organic compounds in water. Most applications of COD determine the amount of organic pollutants found in surface water (e.g. lakes and rivers), making COD a useful measure of water quality. It is expressed in milligrams per liter (mg/L), which indicates the mass of oxygen consumed per liter of solution. Older references may express the units as parts per million (ppm).

The total nitrogen (T-N) in water is comprised of dissolved inorganic and organic nitrogen and particulate organic and inorganic nitrogen, minus N_2 gas. Phytoplankton and bacteria contribute to the amount of dissolved inorganic nitrogen content. Decomposition of aquatic life adds both dissolved organic and particulate organic nitrogen to water; while sewage runoff, erosion, and watershed increases particulate inorganic nitrogen levels in water. Bacterial denitrification converts nitrate to N_2 gas, hence the loss of some of the water's nitrogen.

Total nitrogen consists of inorganic and organic forms. Inorganic forms include nitrate (NO_3^-), nitrite (NO_2^-), unionized ammonia (NH_4), ionized ammonia (NH_3^+), and nitrogen gas (N_2). Amino acids and proteins are naturally-occurring organic forms of nitrogen. All forms of nitrogen are harmless to aquatic organisms except un-ionized ammonia and nitrite, which can be toxic to fish. Nitrite is usually not a problem in water bodies, however, because (if there is enough oxygen available in the water for it to be oxidized) nitrite will be readily converted to nitrate.

T-N in water comes from both natural and man-made sources, including: air (some algae can "fix" nitrogen by pulling it out of the air in its gaseous form and converting it to a form they can use); storm water run-off, including natural run-off from areas where there is no human impact (nitrogen is a naturally-occurring nutrient found in soils and organic matter); fertilizers; animal and human wastes (sewage, dairies, feedlots, etc.).

Total phosphorus (T-P) is a measure of all the various forms of phosphorus that are found in a water sample. Phosphorus is an element that, in its different forms, stimulates the growth of aquatic plants and algae in water bodies. The chemical symbol for the element phosphorus is P and the symbol for total phosphorus is T-P. Some phosphorus compounds are necessary nutrients for the growth of aquatic plants and algae, and some are found naturally in many types of rocks. Mines in Florida and throughout the world provide phosphorus for numerous agricultural and industrial uses.

Socio-economic activities that release wastewater to the environment such as the indexes of number of population including migration and population growth were used for estimation the sources of domestic pollution for each type of settlements. Economic indicators that applied are mainly regional indicators such as gross regional product (GRP). In the basin and zones, all sectors are classifies into 7 categories and GRP, value added, production, input, output, investment for production and treatment, consumption, import, export etc. are mainly used as variables and constrains.

3.3 The Model Specification

3.3.1 Objective Function

Protection and development of the water environment in the upstream parts of Miyun Reservoir are of the greatest priority because this reservoir is the only surface water resource for Beijing. In addition, various trial calculations (the objective function was constructed to minimize T-N, minimize T-P and minimize COD, respectively) revealed that T-N was the pollutant that was the most difficult to degrade. Therefore, an objective function was constructed to minimize the total net load of water pollutant in terms of T-N (superscript 1 represents the water pollutant of T-N) over the target term (t = 10) in order to determine an optimal policy.

$$\text{Min} \sum_{t=1}^{10} TP^1(t) \tag{3.1}$$

We used the data of 2006 for the study area. The basic data were obtained from the Yearbook of Miyun County (2007) and the digital data for the simulations were calculated based on this basic data and related statistics from the State Environmental Protection Administration of China (2002). The dynamic simulation running period was from 2007 to 2016 (10 years).

3.3.2 The Material Flow Balance Model

The net load of water pollutants in the reservoir is defined as the total water pollutants that flow through the rivers, including discharges by sewage plants, and untreated wastes from households, fisheries and rainfall in the catchment area.

3.3.2.1 Water Pollutant Load of Reservoir

The total water pollutant load of reservoir is simulated as the sum of flows through rivers, discharged by sewage plants, fisheries and rainfall on the surface of the reservoir. All the variables are specified in Eqs. (3.3), (3.10)–(3.12).

$$TP^P(t) = \sum_j WP_j^p(t) + FP^p(t) + \sum_m SP_m^p(t) + RP^p(t) \tag{3.2}$$

in which

$TP^P(t)$ the total net load of water pollutant p of the reservoir at time t (t = 1, 2, 3…10) (endogenous);
$WP_j^p(t)$ load of water pollutant p in region j (j = 1 and 2) at time t (endogenous);
$FP^p(t)$ load of water pollutant p by fisheries at time t (endogenous);
$SP_m^p(t)$ water pollutant p discharged by the sewage plant m (endogenous) and
$RP^p(t)$ water pollutant p of the rainfall that flows into the area (exogenous).

Superscript p represents T-N if p = 1, T-P if p = 2, and COD if p = 3.

3.3.2.2 Water Pollutants Flow Through Rivers

All the water pollutant emitted by socio-economic activities flow into the reservoir through revivers.

$$WP_j^p(t) = \kappa_j \cdot SEP_j^p(t) \tag{3.3}$$

in which

κ_j current velocity (exogenous) and
$SEP_j^p(t)$ water pollutant emitted by socio-economic activities that flowing into reservoir through river j (also is region j, j = 1 and 2) at time t (endogenous).

3.3.2.3 Water Pollutant Emitted by Socio-Economic Activities

The total water pollutants emitted by socio-economic activities in the catchment area are composed of pollutants from households, non-point sources and point sources. Pollutants caused by fisheries are specified in Eq. (3.10), because they are directly loaded into the reservoir. Also, pollutants discharged through sewage are specified in Eq. (3.12). All the variables in this formula are specified in Eqs. (3.5)–(3.9).

$$SEP_j^p(t) = HP_j^p(t) + NP_j^p(t) + EFP_j^p(t) \tag{3.4}$$

in which

$HP_j^p(t)$ water pollutant p (excluding those through sewage) loaded by households in region j at time t (endogenous);

$NP_j^p(t)$ water pollutant p emitted by non-point sources in region j (endogenous) and

$EFP_j^p(t)$ water pollutant p emitted by socio-economic activities (excluding those by fisheries) in region j (endogenous).

3.3.2.4   Water Pollutants Caused by Household Wastewater in Each Municipality

Water pollutants caused by household wastewater in each municipality is associated with population of each municipality.

$$HP_j^p(t) = EH^p \cdot P_j^N(t) \tag{3.5}$$

in which

EH^p emission coefficient of water pollutant p by household wastewater without sewage plant (exogenous) and

$P_j^N(t)$ population that does not use sewage plant in region j and discharges pollutants into river (endogenous).

Superscript N represents the population that does not use sewage plant.

3.3.2.5   Load of Water Pollutants Through Non-point Sources

Total load of water pollutants through no-point sources is decided by the land area and coefficient of water pollutant emitted through different land use.

$$NP_j^p(t) = EL^{pk} \cdot L_j^k(t) \tag{3.6}$$

in which

EL^{pk} coefficient of water pollutant p emitted through land use k (k = 1, 2, 3, 4 and 5) (exogenous) and

$L_j^k(t)$ area of land use k in region j that emitted pollutants into river (endogenous).

Superscript k represents the type of land use upland cropping if k = 1, forest land if k = 2, orchard land if k = 3, city area if k = 4, and other land area if k = 5.

3.3.2.6   Load of Water Pollutants by Production Activities

Total load of water pollutants by production activities are composed by six industries pollutants cased by fisheries industry is specifies in Eq. (3.10). In order to evaluate the efficiency of present technology, we set up the Eq. (3.9), especially to show the pollution emitted by pig farming industry.

$$EFP_j^p = EFP_j^{p3\sim6}(t) + EFP_j^{p2}(t) \qquad (3.7)$$

$$EFP_j^{p3\sim6}(t) = EI^{pm} \cdot X_j^m(t)(m \neq 1,2) \qquad (3.8)$$

$$EFP_j^{p2}(t) = \sum_j EI^{pWF} \cdot NA_j^{WF}(t) + \sum_j EI^{pN} \cdot NA_j^N(t) \qquad (3.9)$$

in which

$EFP_j^{p3\sim6}(t)$ water pollutant p emitted by socio-economic activities (excluding those by pig farming and fisheries) in region j (endogenous);

$EFP_j^{p2}(t)$ water pollutant p emitted by stockbreeding industry in region j (endogenous);

EI^{pm} coefficient of water pollutant p emitted by industry m (m = 1, 2, 3...7) (exogenous);

$X_j^m(t)$ production of industry m in the area of region j that emitted pollutants into river (endogenous);

EI^{pWF} coefficient of water pollutant p emitted by pig whose feces and urine treated by wet methane fermentation (exogenous);

$NA_j^{WF}(t)$ number of pig those feces and urine treated by wet methane fermentation in region j (endogenous).

EI^{pN} coefficient of water pollutant p emitted by pig those feces and urine treated without any treatment method (exogenous) and

$NA_j^N(t)$ number of pig those feces and urine treated without any treatment method in region j (endogenous)

Superscript m represents classification of industry, fisheries if m = 1, pig farming if m = 2, other stockbreeding industry if m = 3, forestry if m = 4, planting industry if m = 5, manufacturing industry if m = 6, and other industries if m = 7.

3.3.2.7 Load of Water Pollutants by Fisheries

Total load of water pollutants by fisheries is decided by the production and coefficient of water pollutant emitted by fisheries.

$$FP^p(t) = EF^{pf} \cdot X^f(t) \qquad (3.10)$$

in which

EF^{pf} coefficient of water pollutant p emitted by fisheries (exogenous) and
$X^f(t)$ production of fisheries in the area (endogenous).

3.3.2.8 Load of Water Pollutant of the Rainfall

Total load of water pollutant of the rainfall is given as follows:

$$RP_j^p(t) = ER^p \cdot L^r \tag{3.11}$$

in which

ER^p coefficient of water pollutant p emitted by rainfall (exogenous) and
L^r the catchment's area of the area (endogenous).

3.3.2.9 Water Pollutant Treated by Sewage System

Water pollutant emitted by household is separated by sewage system and without any treatment, and the total load of water pollutant treated by sewage system is associated with the population that uses sewage system.

$$SP^p(t) = ES^p \cdot \sum_j P_j^{SP}(t) \tag{3.12}$$

in which

ES^p the pollution emission coefficient of sewage plant in the area at time t (exogenous) and
$P_j^{SP}(t)$ the population that use sewage plant in region j (exogenous).

Superscript SP represents the sewage plant.

3.3.3 Treatment Policies for Non-point Sources

Miyun government developed the *Organic Fertilizer Promotion Policy* in 2004 and converted the majority of upland cropping into forest land to improve the water conservation function of Miyun Reservoir. This treatment policy is subsidized by the government for the promotion of organic fertilizers.

3.3.3.1 Total Land Area

Total land area is defined as some of five land use areas for each region.

$$\overline{L}_j(t) = \sum_{k=1}^{5} L_j^k(t) \tag{3.13}$$

in which

$\bar{L}_j(t)$ total land area in region j at time t (endogenous) and
$L_j^k(t)$ different types of land use area in region j at time t (endogenous).

3.3.3.2 Conservation of Other Land to Forest Land

Miyun government converted the majority of upland cropping into forest land to improve the water conservation function of Miyun Reservoir.

$$L_j^k(t+1) = L_j^k(t) + \Delta L_j^k(t) \tag{3.14}$$

in which
$\Delta L_j^k(t)$ increase in forest land that conversed from other land use in region j at time t (endogenous) and $k = 2$.

$$\Delta L_j^k(t) = L_j^{52}(t) \tag{3.15}$$

in which
$L_j^{52}(t)$ conversion of land use from other land (superscript $= 5$) to forest land (superscript $= 2$) in region j at time t (endogenous) and $k = 2$.

The conversion from other land area to forest land is subsidized by the Miyun government.

$$L_j^{52}(t) \geq \lambda^5 \cdot S_j^{5f}(t) \tag{3.16}$$

in which

$L_j^{52}(t)$ conversion of land use from other land area to forest land in region j at time t (endogenous);
λ^5 reciprocal of the subsidy for one unit conversion to forest (exogenous) and
$S_j^{5f}(t)$ subsidy given by the government for conversion of land use (endogenous).

The change rate of variable $L_j^k(t)$ was indicated as endogenous variable $\Delta L_j^k(t)$ to indicate how the variable changes from year t to t + 1. The change rate was constrained and depended on the value of endogenous variable $S_j^{5f}(t)$ which can be achieved through automatic calculation process (looping computation and iterations) based on the interactive and restrictive correlations among variables.

3.3.3.3 Conservation of Land Area with Organic Fertilizer

Miyun government developed the *Organic Fertilizer Promotion Policy* in 2004. This treatment policy is subsidized by the government for the promotion of organic fertilizers.

$$L_j^1(t) = LC_j^{IF}(t) + LC_j^{OF}(t) \tag{3.17}$$

in which

$L_j^1(t)$ the total area of planting in region j at time t (endogenous);
$Lc_j^{IF}(t)$ the area of planting which applied with inorganic fertilizer in region j at time t (endogenous) and
$LC_j^{OF}(t)$ the area of planting which applied with organic fertilizer in region j at time t (endogenous).

Superscript IF represents inorganic fertilizer and OF represents organic fertilizer.

$$LC_j^{OF}(t+1) = LC_j^{OF}(t) + \Delta LC_j^{OF}(t) \tag{3.18}$$

in which
$\Delta LC_j^{OF}(t)$ increase in planting area which applied with organic fertilizer in region j at time t (endogenous).

The increase in planting area which applied with organic fertilizer is decided by the subsidization provided by the Miyun government.

$$\Delta LC_j^{OF}(t) = ILC_j^{OF} \cdot S_j^{OF}(t) \tag{3.19}$$

in which

ILC_j^{OF} increasing amount of planting area which applied with organic fertilizer per thousand RMB Yuan in region j at time t (exogenous);
$S_j^{OF}(t)$ subsidization for the utilization of organic fertilizer in region j at time t (endogenous).

3.3.4 Policies for Household Wastewater Generation

According to the characteristics and situations of suburbs of big cities, the treatment of household wastewater mainly depends on the sewage systems and sewage plants. These measures are implemented by the local governments.

3.3.4.1 The Change of Population

The change of population is given as follows:

$$P_i(t+1) = P_i(t) + \Delta P_i(t) \tag{3.20}$$

in which

$P_i(t)$ population of municipality i (i = 1, 2, 3...9) at time t (endogenous) and
$\Delta P_i(t)$ increase in the population of municipality i at time t (endogenous).

Superscript i represents the index of nine towns in the study area.

The population can be separated as that use sewage plant and receive without any wastewater treatment. According to this formula, the Eqs. (3.5) and (3.12) can be implemented.

$$P_i(t) = P_i^{SP}(t) + P_i^{N}(t) \tag{3.21}$$

in which

$P_i^{SP}(t)$ the population that use sewage plant of municipality i (endogenous) and
$P_i^{N}(t)$ the population that without any wastewater treatment of municipality i (endogenous).

$$\Delta P_i(t) = \Delta P_i^{SP}(t) + \Delta P_i^{N}(t) \tag{3.22}$$

in which

$\Delta P_i^{SP}(t)$ increase in the population that use the sewage plant (endogenous) and
$\Delta P_i^{N}(t)$ increase in the population that without any wastewater treatment (endogenous).

3.3.4.2   Increase in the Population that Uses the Sewage System and Plant

The increase in the population that used the sewage system and plant are dependent on the construction investment which is specified by Eqs. (3.25)–(3.28).

$$\Delta P_i^{SP}(t) \le \lambda_i^{SP} \cdot I_i^{SP}(t) \tag{3.23}$$

in which

λ_i^{SP} reciprocal of the necessary construction investment per person that uses the sewage plant (exogenous) and
$I_i^{SP}(t)$ construction investment of municipality i for sewage plant (endogenous).

3.3.4.3   Local Finance

Majority of construction investment and maintenance cost of sewage plant are dependent on the local finance of each municipality.

$$FE_i(t+1) = FE_i(t).(1 + \theta_i) \tag{3.24}$$

in which

$FE_i(t)$ total finance of municipality i at time t (endogenous) and
θ_i the rate of total finance growth of municipality i at time t (exogenous).

3.3.4.4 Sewage System

Investments for construction of sewage systems and plants are determined by the construction allotment of the municipality and subsidies that are provided by the local government. The construction allotment and maintenance costs are covered by local finances and special subsidizations.

$$K_i^{SP}(t+1) = K_i^{SP}(t) + I_i^{SP}(t) \tag{3.25}$$

$$I_i^{SP}(t) = \left(\frac{1}{1 - M_i^{SP}}\right) \cdot \xi_i^{SP} \cdot FE_i(t) \tag{3.26}$$

in which

$K_i^{SP}(t)$ capital available for sewage system in municipality i at time t (endogenous);

M_i^{SP} rate for investment of sewage system from the municipality i (exogenous) and

ξ_i^{SP} rate transfer for investment of household wastewater from the total finance of municipality i (exogenous).

     The maintenance cost of sewage system is mainly provided by local finance. The total investment and maintenance cost for construction of sewage system are dependent on the local finance and household wastewater treatment subsidy that granted by the Miyun government to the municipality for intensive promotion of construction and installation for treatment measures of household wastewater.

$$MC_i^{SP}(t) = \varsigma_i^{SP} \cdot FE_i(t) \tag{3.27}$$

$$I_i^{SP}(t) + MC_i^{SP}(t) \leq \varpi \cdot FE_i(t) + S_i^{SP}(t) \tag{3.28}$$

in which

MC_i^{SP} maintenance of sewage system of municipality i at time t (endogenous);

ς_i^{SP} rate transfer for maintenance cost of household wastewater from the total finance of municipality i (exogenous);

ϖ rate transfer for treatment of household wastewater from the total finance of municipality i (exogenous) and

$S_i^{SP}(t)$ subsidy for households wastewater treatment in municipality i, that is granted by Miyun government (endogenous).

3.3.5 Treatment Measures for Production Generation Sources

3.3.5.1 Production Function and Curtailment

This production function is derived from Harrod–Domar model through the relationship between capital accumulation and production. We assumed the production of industry m is restricted by leaving capital idle and subsidy for loss due to the idle capital.

$$X_j^m(t) \le \alpha^m \cdot \left\{ K_j^m(t) - S_j^m(t) \right\} \tag{3.29}$$

in which

α^m ratio of capital to output in industry m (exogenous) and
$S_j^m(t)$ subsidy for industry m in region j, that is granted by Miyun government (endogenous).

The capital accumulation is dependent on the investment and depreciation of capital.

$$K_j^{mp}(t+1) = K_j^{mp}(t) + I_j^{mp}(t+1) - d^m \cdot K_j^{mp}(t) \tag{3.30}$$

in which

$K_j^{mp}(t)$ capital available for industry m in region j at time t (endogenous);
$I_j^{mp}(t)$ investment in industry m in region j at time t (endogenous) and
d^m depreciation rate of industry m (exogenous).

3.3.6 Total Budget of the Government for the Countermeasures

It is assumed that the Miyun government spends 149.3 million RMB Yuan for implementing the countermeasures every year. This figure is based on the actual budget that has been directly and indirectly spent to improve the quality of the reservoir. These variables are specified in Eqs. (3.16), (3.19), (3.28) and (3.29).

$$y(t) \ge \sum_i S_i^{SP}(t) + \sum_j \sum_m S_j^m(t) + \sum_j S_j^{OF}(t) + \sum_j S_j^{5f}(t) \tag{3.31}$$

in which
$y(t)$ the total budget spent by the local government for implementing the countermeasures (exogenous).

3.3.7 Flow Balance in the Commodity Market

Each industry must produce in order to meet the balance between supply and demand for the commodity produced in the industry. The production is dependent on Leontief input–output coefficient matrix, consumption, investment and net export. The total consumption is specified in Eq. (3.34), investment of industry and sewage plant are specified in Eqs. (3.30) and (3.26), the net export represented as Eq. (3.33).

$$X(t) \geq A \cdot X(t)C(t) + i^m(t) + B^{SP} \cdot I^{sp} + e(t) \tag{3.32}$$

in which

$X(t) = \sum_j X_j(t)$	column vector of the mth element that is the total product of industry m in the basin (endogenous);
A	input–output coefficient matrix (exogenous);
$C(t)$	the total consumption at time t (endogenous);
$i^m(t) = \sum_j I_j^{mP}(t)$	the total investment at time t (exogenous);
B^{SP}	column vector of mth coefficient that induced production in industry m by construction of sewage plant (exogenous);
I^{SP}	total investment for construction of sewage plant (endogenous) and
$e(t)$	column vector of net export (endogenous).

3.3.8 Restriction of Net Export

The restriction of net export is given as follows:

$$e_{min} \leq e(t) \leq e_{max} \tag{3.33}$$

in which

$e_{min}(t)$ column vector of minimum net export (exogenous) and
$e_{max}(t)$ column vector of maximum net export (exogenous).

3.3.9 Restriction of Consumption

The restriction of consumption is dependent on the population and column vector of the consumption of each industry.

$$C(t) \geq H \cdot \sum_i P_i(t) \tag{3.34}$$

in which
H column vector of the mth element that is the consumption of industry m in the basin (exogenous).

3.3.10 Gross Regional Product

We adopt the index of GRP to reflect the development of the local socio economy, which is dependent on the production and added value of each industry.

$$GRP(t) = \upsilon \cdot X(t) \tag{3.35}$$

in which

$GRP(t)$ gross regional product (endogenous) and
υ row vector of mth element that is rate of added value in mth industry
          (exogenous).

3.3.11 Constraints on the Regional Economy

It is assumed that the GRP realize annual increase or at least equal the GRP of last
year. This formula is based on the actual characteristics and requirements of economic development in China.

$$GRP(t + 1) \geq GRP(t) \tag{3.36}$$

3.4 Simulation Results

We defined the simulation cases that reflected the level of economic development
(gross regional product, GRP) as increases of n % per year (annual growth rate of GRP)
during the future 10 years in the study area. Four representative cases were selected to
demonstrate the variations in the simulation results. These cases were numbered so as
to facilitate the simulation analysis. When a GRP increase of 0 % per year, we set the
name as Case 1 to represent the minimum level of economic development. When GRP
realize an increase of 0.7 % per year, we set the name as Case 2. And Case 3 shows an
increase of 1 % per year. Case 2 and Case 3 were selected to express the tendency of
simulation results. When a GRP increase of 1.1 % per year, we set the name as Case 4
to indicate the maximum level of economic development in this simulation.

3.4.1 Objective Function

In this basic model, we adopted integrated policies for the simulation. However,
with strict restrictions on the objective function to minimize T-N, the annual
growth rate of GRP only reached 1.1 % with the global solution and no feasible
solution could be obtained for greater than 1.1 %. The feasible optimal solution
was achieved in Case 4 when we adopted the integrated polices.

Comparing the results of simulation, the introduction of pollutant-minimization
policies is effective to reduce water pollutants. When introduction of the optimal
pollutant-minimization policies and GRP increased by 1.1 % per year (Case 4), the
objective value of T-N (sum of T-N for 10 years) decreased about 10.7 % as compared to when not adopted (Fig. 3.2). Moreover, with the integrated policies, the
GRP had a difference of about 192.5 million RMB as compared to the initial year.

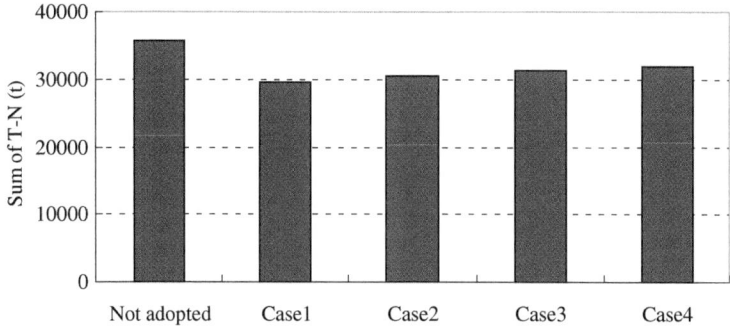

Fig. 3.2 Objective value for total nitrogen (T-N)

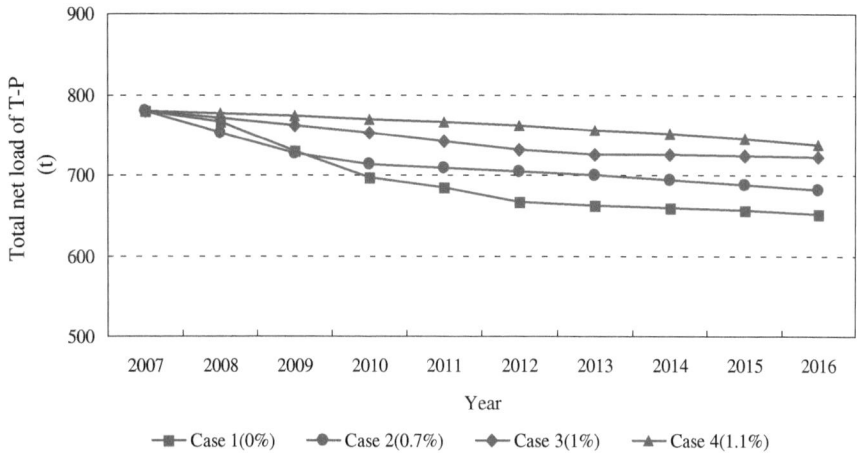

Fig. 3.3 Changes in T-P

The variations in total amounts of T-P and COD flowing into the Miyun Reservoir are shown in Figs. 3.3 and 3.4. In Case 4, the net load of T-P and COD was reduced 14.2 and 19.0 %, respectively, in 2016 with the integrated polices when compared with no adoption of the policies. This result verifies that it is necessary to consider the minimization of T-N as the objective function when we formulate integrated policies to improve the water environment.

3.4.2 Changes in GRP

The annual growth rate of GRP only reached 1.1 % with the objective function of water pollutants. Compared to the initial year, GRP increased 9.88 % in Case 4 (Fig. 3.5). We conclude that as the optimal solution of the simulation, an annual growth rate of GRP could only reach 1.1 % with the restrictions on the water

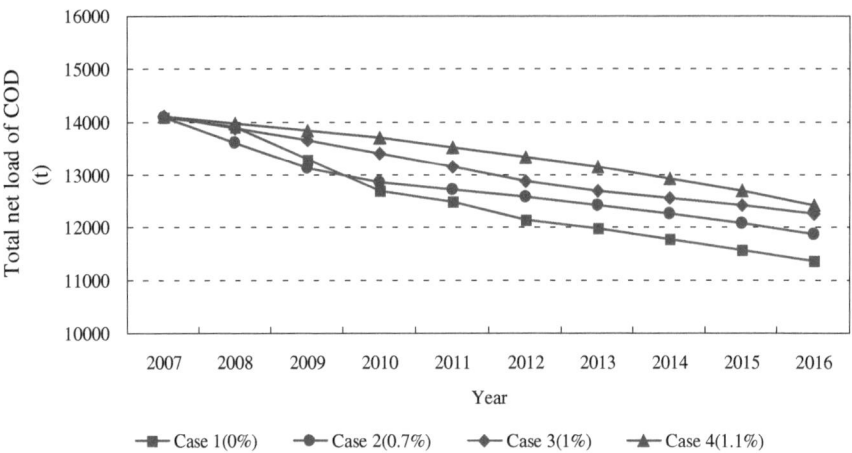

Fig. 3.4   Changes in COD

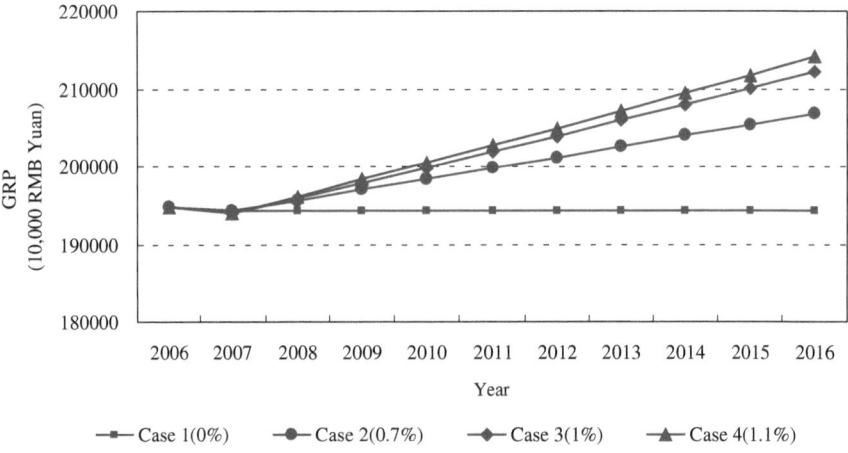

Fig. 3.5   Changes in gross regional product (GRP)

environment. The economic growth rate is far lower than the average level of the suburbs of big cities in China, and even below the economic target of the catchment area (annual growth rate maintained at 5–6 %).

3.4.3   Variations in Production and Water Pollutants Flowing into Reservoir from Pig Farming

The variations of production and water pollutants (T-N) from the pig farming industry that flow into the reservoir are shown in Fig. 3.6 and Table 3.2. The production of pig farming decreased more than 30 % as compared to 2006. Moreover, the

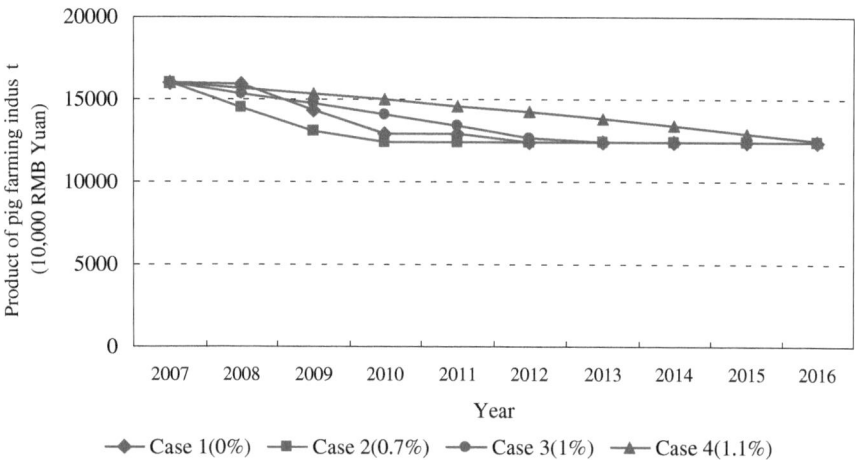

Fig. 3.6 Changes in production of pig farming industry

T-N	Case 4 (GRP increase of 1.1 %/year)	
	Region 1	Region 2
2007	215.66	595.35
2008	211.34	583.41
2009	206.80	570.90
2010	202.06	557.79
2011	197.09	544.07
2012	191.89	529.72
2013	186.46	514.73
2014	180.79	499.09
2015	174.88	482.76
2016	168.71	465.75
Total	1,935.67	5,343.57

Table 3.5 Emissions of T-N from pig farming industry flowing into the reservoir (Case 4) Unit: t

water pollutants from the pig farming industry that flow into the reservoir were only slightly reduced even with the adoption of wet methane fermentation technology. Based on the results in Table 3.5, the towns of Region 2 emit 2.76 times more total nitrogen into the catchment area of the Miyun reservoir than Region 1 in the basin.

The following reasons are enumerated for these results. First, the pig farming industry has the largest water pollutants emission coefficient, especially T-N. Second, with the requirements of Miyun Reservoir and objective function, serious generation of pollution restricts the development of the pig farming industry. This result shows that the contribution to water pollution by wastewater from the pig farming industry is higher than considered and the efficiency of wet methane fermentation technology is not sufficient to treat the high concentrations of wastes in the water. In addition, the Miyun government has raised a strategy for "Strong Stockbreeding Industry" as a regional development orientation in order to improve

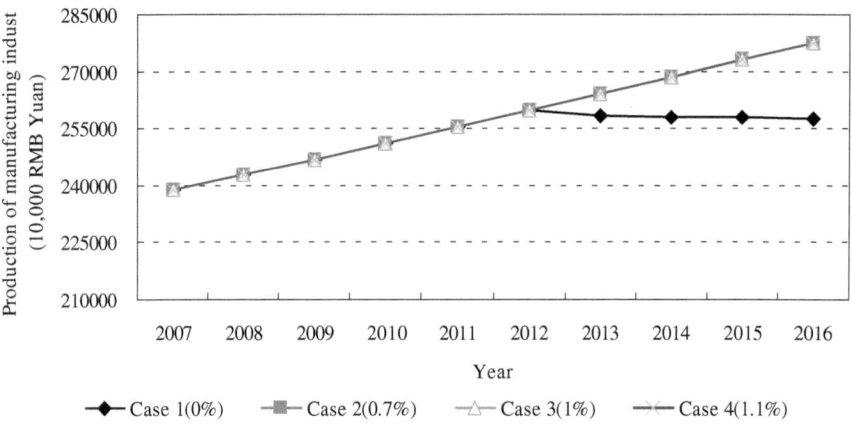

Fig. 3.7   Changes in production of manufacturing

regional economic development and satisfy the increased demand for animal products in Beijing. It is impossible to reduce the pig farming production in reality.

   Therefore, it is necessary and essential to introduce advanced technologies to
treat the serious pollution problems and improve regional economy. The most
important factor for the development of advanced technologies in the future is
to enable treatment of wastewater from pig farming for higher concentrations of
pollutants.

3.4.4 Production of Manufacturing Industry

Variations in manufacturing industry production in each case are shown in Fig. 3.7.
The production of the manufacturing industry steadily increased for 6 years in Case
1, and dramatically increased in Cases 2, 3 and 4. China is under industrialization, the
manufacturing industry plays a dominant role in most cities and the suburbs of big
cities. Therefore, the production of manufacturing industry steadily increases, even
in Case 1. This simulation verifies the distinguishing features and the actual situation
of China. On the other hand, the results show that the total amount of production of
the manufacturing industry in Case 2 and Case 3 are equivalent to Case 4. Therefore,
we can conclude that with restrictions on the environment and the low efficiencies in
treatment technology, it is difficult to greatly improve economic development.

   In addition, we also find that the production of other stockbreeding industries
increases year by year (Case 4) with the adoption of wet methane fermentation
technology, increased subsidies for water conservation, and water conservation
forests increasing to about 0.75 km^2. Therefore, we concluded that the integrated
polices are effective to control water pollutants emitted by the sources in the
objective areas except for pig farming.

3.5 Conclusions and Discussions

In this study, we adopted the integrated policies to minimize water pollutants flowing into Miyun Reservoir based on two sub-models. Particularly, we improved the economic models so that they are more adaptive to the specific characteristics and actual situation of suburbs of big cities in China. The following results were obtained by this simulation. First, when the integrated pollutant-minimization policies with wet methane fermentation technology were introduced, GRP increased 1.1 % per year, and the objective value of T-N decreased about 10.7 % as compared to when the policies were not adopted. Second, with the integrated policies, GRP had a difference of about 192.5 million RMB as compared to the initial year. Third, we found that 266.7 t of T-N, 105.1 t of T-P and 1,641.1 t of COD emitted by the pig farming industry in 2016 can be reduced as compared to initial year by introducing the integrated policies.

Based on the results of the simulation, the adoption of wet methane fermentation technology reduced a fraction of water pollutants emitted by the pig farming industry in the catchment area. However, the limited environmental preservation is possible at the cost of significant reduction of the production of pig farming industry (the production decreases 30 %) and slow-growing regional economy (annual growth rate of economy only reach 1.1 %). In addition, there is security problem of the sealing device in the application of wet methane fermentation technology in China. On the other hand, with increased demand for animal products in Beijing and the development of a "Strong Stockbreeding County" by the Miyun government, there is no reason to think that we must deteriorate the socio-economic activity level to improve the quality of environment. Therefore, the efficiency of wet methane fermentation technology is not sufficient to treat the high concentrations of wastes in the water and meet the requirements of future economic development in the catchment area. Especially, the pig farming industry requires the introduction of advanced technology to allow simultaneous pursuit of environmental preservation and economic development. The simulation results also verify this comprehensive model is adaptive to specific situations in China and conclusions are general applicable to the suburbs of big cities in China.

While the potential of this application has been demonstrated in the case of water pollutant-minimization problem, the synthetic considerations for ecologic system evaluation need further exploration. We also should consider reducing the abundance of greenhouse gases (GHGs) over the whole ecosystem by evaluating changes in material balances. In addition, in this study, it was found that the efficiency of wet methane fermentation technology is not sufficient to treat the wastewater with high concentrations of pollutants, and that the most important factor for the technologies is how enable treatment of wastewater with higher concentration, especially the water pollutant T-N. Moreover, stockbreeding waste is a carbon-neutral biomass resource that can be used to produce biomass energy through advanced technologies. Therefore, in formulating synthetic policies to reduce the quantities of water pollutants emitted by the stockbreeding industry, we should consider the comprehensive use of livestock feces and urine as biomass resources. These projects are expected to help improve biomass resource utilization and environmental protection and the basis of decision-making for sustainable development in the rural areas surrounding China's large cities.

References

Freeman AM, Havemen RH, Kneese AV (1973) The economics of environmental policy. American, p 175

Hadley G (1962) Linear programming. Addison-Wesley Press, New York, pp 10–54

Hillier FS, Gerald JL (1967) Introduction to operations research. Holden-Day, Inc. Press, San Francisco, pp 107–121

Wagner HM (1959) Linear programming techniques for regression analysis. J Am Stat Assoc 57:206–212

Yan JJ, Xu F, Higano Y (2010) Comprehensive evaluation of integrated pollutant-minimization policies in rural area around Beijing: case study of Miyun County. J Hum Environ Symbiosis 17:87–98

Chapter 4
Analysis of Optimal Environmental Policies with Advanced Technologies for Effective Treatment of Stockbreeding Wastes

Abstract This Chapter consists (1) a literature review of the main environmental policies regarding reduction of pollution and appraisal of stockbreeding wastes treatment methods; (2) the introduction of integrated policies framework that were raised to realize effective utilization of stockbreeding wastes and environmental preservation; and (3) the introducing of advanced technologies for proper treatment of stockbreeding wastes in the study area.

Keywords Environmental policies · Integrated policies · Advanced technologies

4.1 Concepts of Integrated Environmental Policies

4.1.1 Direct Regulation

It is a traditional environmental policy. The government enacts a series of environmental standards that are to be followed. Direct control can be of two wide types: technology-based and performance-based regulations. The first type specifies the methods and equipment firms should use to meet the standard while the second set an overall target for each enterprise and give them some direction in how to meet the standard (Ikkatai 1996).

However, this approach is being criticized due to: (1) monitoring is expensive and sometimes difficult (Frijins and Vliet 1999; William and Wallace 1979); (2) ignores the ability of some industries especially small-scale and micro-scale, to reduce their pollution. This may force many of the firms to close down. This is common in developing countries (O'Connor 1999); (3) this approach does not give any incentives for firms to reduce their emissions beyond the standard (OECD 1997).

© The Author(s) 2015
J. Yan, *Comprehensive Evaluation of Effective Biomass Resource Utilization and Optimal Environmental Policies*, SpringerBriefs in Economics,
DOI 10.1007/978-3-662-44454-2_4

4.1.2   Economic Instruments

The aim of economic instruments is to charge the polluters for their damage to the environment (Frijns and Vliet 1999). This approach encourages the polluter to decrease the pollution as much as possible to avoid paying for the damage and gives them a chance to choose the best suitable for them to reduce their emission (OECD 1991). Considering both, environmental protection and economic development, direct control is doubted for its economic efficiency. Instead, economic instruments including emission charges or environmental tax, subsidies, deposit-refund systems and discharge permit trading have drawn attentions for its use to pollution control (Pearce et al. 1999; OECD 1994).

Emission charge is a fee, collected by the government, levied on the each unit of pollutant emitted to air or water. In general, emission charges reduce pollution because firms seek ways to reduce cost by reducing pollution (OECD 1997). The firm would select to reduce emissions until the marginal cost of reduction is equal to the emission charge rate. These charges can be user charges so only those who are connected to the public services are charged. In addition, product charges are applied for raw material, intermediate or final products.

Subsidies are another form of economic instruments although then can lead to economically inefficient situations. However, then are used in certain circumstances, such as payment for positive externalities and supporting firms for using green technologies (OECD 1991).

The basic ideas of the tradable permit are (1) local government decides the total amount of pollution to be permitted in an area; (2) the total pollution discharges permits are allocated to each existing dischargers by some rules; (3) the holders of these permits can be traded as a free market commodity. The permit market induces cost efficiency by encouraging dischargers capable of more cost efficient pollutant reduction to operate at high removal levels and to hold few permits while allowing less efficient dischargers to operate at relatively low treatment levels and to hold larger number of permits (Baumol and Oates 1988; Atkinson and Stern 1974).

In deposit-refund systems a deposit is paid on potentially polluting products. When pollution is avoided by returning the products or their residuals, a refund follows as a reward of good environmental behavior (OECD 1994).

Comparison among the policies of command and control, emission charges, and tradable permits is quite controversial. Direct regulation is environmentally effective. However, discontinuous control might negatively affect the cost efficiency of standards of the equipment has to be operated at less than full capacity. Emission charges and tradable permits are clearly more cost efficient and have more advantages in creating higher incentives. However, in and imperfect enforcement and an imperfect market where there exists uncertainty about costs, inflation has a negative impact on the environmental efficiency of charges and the cost efficiency on cost efficiency for permits (Klaassen 1996).

4.1.3 Integrated Approach

Combinations of economic instruments and direct regulations are quite common and fruitful. For instance, emission charges are combined with direct regulation, either to reinforce such regulations or to provide necessary funds. Taxation on polluting products and subsidies on cleaner is another successful combination (OECD 1991).

Yet even where economic instruments are successfully implemented, there is always a role for regulation. For example, where flexibility of response cannot be allowed or where is imperative that the emissions of certain toxic pollutants are prohibited, then the government relies on regulation to enforce compliance (OECD 1997).

4.2 Integrated Policy Framework

4.2.1 Classification of Pollutants and Sources

The classification of sub-basins of the main rivers is the same as Chap. 2. The pollutants measured in this study included three water pollutants, T-N (total nitrogen), T-P (total phosphorus) and COD (chemical oxygen demand) and three greenhouse gases, CO_2, CH_4 and N_2O (Tables 4.1 and 4.2). CH_4 and N_2O have a much greater greenhouse effect potential than CO_2. Therefore, to appropriately evaluate emission of greenhouse gas, we have to consider the greenhouse effect potential of each gas (see Fig. 4.1). The household wastewater disposal system was divided into two categories based on the situations in the rural suburbs, land use into 5 categories of non-point generation sources and industry into 7 categories of production generation based sources on the characteristics of Miyun County (Tables 4.3, 4.4 and 4.5).

Table 4.1 Classifications of water pollution parameter

Index	Parameter
1	T-N
2	T-P
3	COD

Table 4.2 Classifications of green house gases parameter

Index	Parameter
1	CO_2
2	CH_4
3	N_2O

Fig. 4.1 Potential index of greenhouse effect. *Source* IPCC (2006), pp 212

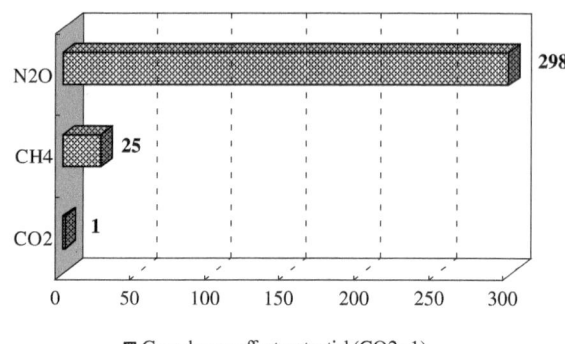

☒ Greenhouse effect potential (CO2=1)

Table 4.3 Classifications of household waste water disposal system

Index	Facility
1	Sewage plant
2	Untreated waste water

Table 4.4 Classifications of land use

Index	Land use
1	Upland cropping
2	Forest land
3	Orchard land
4	City area
5	Other land area

Table 4.5 Classifications of industry

Index	Industry
1	Fisheries
2	Pig farming
3	Other stockbreeding industries
4	Forestry
5	Planting industry
6	Manufacturing industry
7	Other industries

4.2.2 Integrated Policies

Table 4.6 and Fig. 4.2 represent the integrated policies used in the simulation model to reduce water pollutants and greenhouse gases, and the government selected advanced policies to realize efficient improvement of the regional

Table 4.6 Integrated policies for the catchment area

Source of pollutants	Name of policy	Index	Measurements
Production	Reduction of the capital employed	1	Subsidization for industries to reduce working capital so as to adjust production
Pig farming industry	Introduction of advanced technologies (*biomass recycle plant* and *new energy project*) and reduction of the capital employed	2	Subsidization for pig farming industry to introduce *biomass recycle plant*
		3	Subsidization for pig farming industry to introduce *new energy project*
		4	Subsidization to reduce working capital so as to adjust production
Household	Sewage system and sewage plant	5	Subsidization for the municipality to install more sewage system and sewage plant
Non-point	Promotion of organic fertilizer	6	Subsidization for farmer who uses organic fertilizer
Water conservation	Water conservation forest	7	Subsidy to promote water conservation forest (conservation of other land to forest)

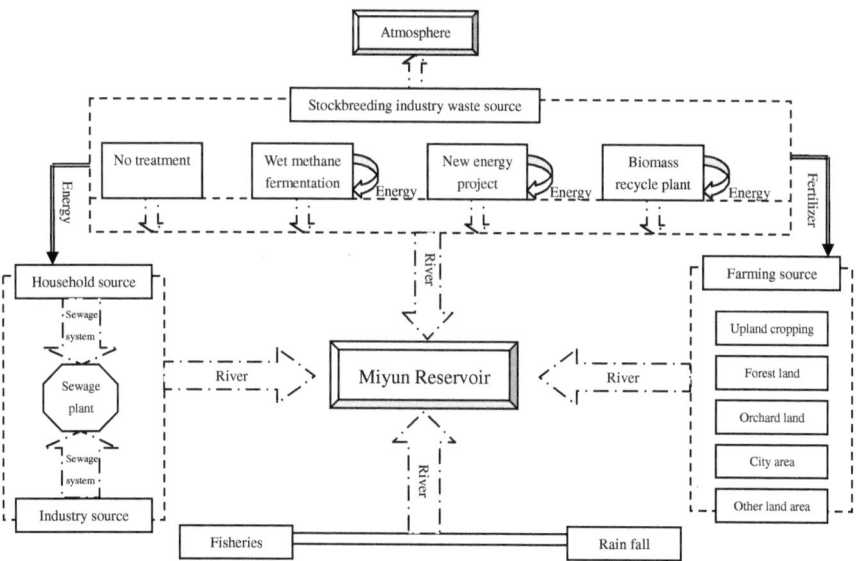

Fig. 4.2 Framework of the simulation model

environment. According to the results of Chap. 3, pig farming industry has become the source of the most serious pollutants to Miyun Reservoir. More than 25 % of T-N, 41 % of T-P and 36 % of COD come from the pig farming industry. The catchment area has introduced wet methane fermentation technology to treat serious water pollutants emitted by the stockbreeding industry, particularly pig farming. However, the efficiency of wet methane fermentation technology is not sufficient to treat the high concentrations of wastes in the water and meet the requirements of future economic development in the catchment area. Therefore, in this part we raised integrated policies with introduction of two advanced technologies (*New Energy Project* and *Biomass Recycle Plant*) to particularly treat serious pollutants emitted by the pig farming industry.

The government proposed and provided preferential prices for farmers to promote utilization of organic fertilizers. The annual budget that the government spends directly and indirectly for the improvements of the Miyun environment is approximately 149.3 million RMB Yuan. The hypotheses are the same with that in Chap. 3.

4.3 Advanced Technologies for Proper Treatment of Stockbreeding Wastes

4.3.1 Introduction of Technologies

A *Biomass Recycle Plant* was developed by Ibaraki Preference as a city area project for promotion of coordination between industry, academia and government, and Advancement of Regional Science and Technology: "The Lake Ksumigaura Biomass Recycling Development Project" (Mizunoya et al. 2006). The process consists of methane fermentation, electrified treatment and heat and power supply system.

A *New Energy Project* was developed by Hangzhou City of China to promote effective and comprehensive utilization of pig wastes, and the project was classified as a state-level pilot project. The process consists of anaerobic fermentation, anaerobic-aerobic combined treatment and biogas-diesel dual fuel for power generation.

4.3.2 Efficiency of the Advanced Technologies

The pollutant efficiency and installation cost of the two advanced technologies is shown in Tables 4.7 and 4.8. Comparing with the general treatment method for stockbreeding wastes, the treatment efficiencies of two advanced technologies are much higher, specially the technology of *Biomass Recycle Plant* (*Biomass plant*).

Table 4.7 Pollutants coefficient of two advanced technologies

Pollutant kg/one pig-year	Pile up outdoors	Compost	New energy project	Biomass plant
CO_2	4	4	35.74	123.09
	(carbon neutral)	(carbon neutral)	(carbon neutral)	(carbon neutral)
CH_4	5.38	2.69	0	0
N_2O	0.0986	0.611	0.00069	0.0005
T-N	7.01	5.18	0.8363	0.034
T-P	0.047	0.047	0.0138	0.003
COD	3.8	2.85	1.0475	0.034

Source (Higano et al. 2006, pp 238–326; Hangzhou Energy and Environment Engineering Co., Ltd. (HEEE) (2003), pp 10–53)

Table 4.8 Installation cost of advanced technologies

Data	New energy project	Biomass recycle plant	Unit
Construction cost	2.81 million	6.5 million	RMB Yuan
Maintenance cost	0.22 million	0.39 million	RMB Yuan
Electricity generated	209.88 thousand	227.59 thousand	kWh/year
Target scale of pig farm	10,000	1,000	Pigs

Source (Higano et al. 2006, pp 238–326; Hangzhou Energy and Environment Engineering Co., Ltd. (HEEE) 2003, pp 10–53)

However, the installation costs of two technologies are more expensive than general treatment methods. Therefore, we should search the optimal arrangement of two advanced technologies based on the restriction of subsidy and requirement of environment in different towns of the study area.

References

Atkinson AB, Stern NH (1974) Pigou, taxation and public goods. Rev Econ Stud 41(1):119–128 Blackwell Publishing

Baumol WJ, Oates WE (1988) The theory of environmental policy. Cambridge University Press, UK, pp 156–190

Frijns J, Vliet BV (1999) Small-scale industry and cleaner production strategies. World Devel 27(6):967–983 Elsevier Science Ltd.

Hangzhou Energy and Environment Engineering Co., Ltd (HEEE) (2003) New energy project report, pp 10–53

Higano Y, Mizunoya T, Piao SH (2006) Ibaraki preference in the city area project for promotion of coordination between industry, academia and government, and advancement of regional science and technology: the lake Ksumigaura biomass recycling development project (2003–2005), pp 238–326

Ikkatai S (1996) The latest state of the water quality management policy. J Res Environ 34(3):5–10

IPCC (2006)IPCC guidelines for national greenhouse gas inventories http://www.ipcc-nggip.iges. or.jp/public/2006gl/index.html pp 212

Klaassen G (1996) Acid rain and environmental degradation: the economics of emission trading. UK, Edward Elgar, Cheltenham, pp 58–94

Mizunoya T, Sakurai K, Kobayashi S, Piao SH, Higano Y (2006) A simulation analysis of synthetic environment policy: effective utilization of biomass resources and reduction of environmental burdens in Kasumigaura basin. Stud Reg Sci 36(2):355–374

O'Connor D (1999) Applying economic instruments in developing countries: from theory to implementation. OECD Development Centre, Paris, pp 65

OECD (1991) Environmental policy: how to apply economic instruments, organization for economic co-operation and development documents. Paris, pp 130

OECD (1994) Managing the environment: the role of economic instruments, organization for economic co-operation and development documents. Paris, pp 191

OECD (1997) Evaluating economic instruments for environmental policy, organization for economic co-operation and development documents. Paris, pp 141

Pearce D, Markandya A, Barbier EB (1999) Blueprint of a green economy. Earthscan Publications Ltd, London, pp 172

William JB, Wallace EO (1979) Economics, environmental policy and the quality of life. Prentice-Hall Inc, Englewood Cliffs, USA, pp 368

Yan JJ, Sha JH, Chu X, Xu F, Higano Y (2014) Endogenous derivation of optimal environmental policies for proper treatment of stockbreeding wastes in the upstream region of the Miyun Reservoir, Beijing. Pap Reg Sci 93:477–500

Chapter 5
Comprehensive Evaluation of Environmental Policies with Advanced Technologies for Effective Utilization of Biomass Resource

Abstract In this Chapter, we improved and modified the simulation model in Chap. 3. In the simulation, we considered the specific and special characteristics of study area's economy and social state in terms of sustained economic growth rate, financial subordination relations, and regional environmental policies, which are different with the model of Japan. Besides these, we introduced two different advanced technologies from Japan and China to the study area through simulation with integrated policies and carried out regional analysis and allocation for two technologies. The chosen integrated policies in Chap. 4 are introduced as endogenous variables in the model in order to do comprehensive evaluation of environmental policies with advanced technologies for effective utilization of stockbreeding wastes.

Keywords Comprehensive evaluation · Environmental policies · Integrated policies · Advanced technologies · Biomass resource

5.1 Model Structure and Characteristics

This dynamic simulation model consisted of a water pollutants flow balance model, air pollutants balance model, energy balance model, socio-economic model and one objective function. Figure 5.1 illustrated model concept. The advanced policy was derived to minimize T-N, which was subjected to sub-models. The water pollutants flow balance model describes how the pollutants flow into the rivers and reservoir. The air pollutants flow balance model describes how the air pollutants are emitted in the objective area. The energy balance model simulated how much biomass energy can be produced by the two advanced technologies. The socio-economic model represents the social and economic activities in the region and the relationship between the activities and emission of pollutants.

© The Author(s) 2015
J. Yan, *Comprehensive Evaluation of Effective Biomass Resource Utilization and Optimal Environmental Policies*, SpringerBriefs in Economics,
DOI 10.1007/978-3-662-44454-2_5

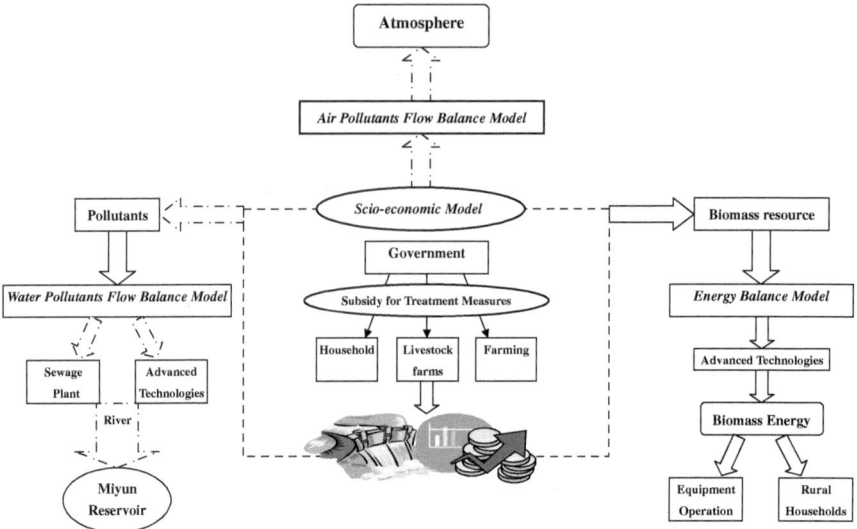

Fig. 5.1 Model concept

5.1.1 Objective Function

The protection and development of the water environment holds the highest priority in the upper stream area of the reservoir because Miyun Reservoir is the only surface water resource for Beijing. In addition, we investigated T-N as the most difficult degradable pollutant through different trial calculations. Therefore, an objective function was constructed to minimize the total net load of water pollutant in terms of T-N over the target term (t = 10) in order to determine an optimal policy.

$$\text{Min} \sum_{t=1}^{10} TP^1(t) \tag{5.1}$$

5.1.2 Water Pollutants Flow Balance Model

The net load of water pollutants in the reservoir is defined as the total water pollutants that flow through the rivers, including discharges by sewage plants, and untreated wastes from households, fisheries and rainfall in the catchment area.

5.1.2.1 Water Pollutant Load of Reservoir

The total water pollutant load of reservoir is simulated as the sum of flows through rivers, discharged by sewage plants, fisheries and rainfall on the surface of the reservoir. All the variables are specified in Eqs. (5.3), (5.10), (5.12) and (5.1).

$$TP^p(t) = \sum_j WP_j^p(t) + FP^p(t) + \sum_m SP_m^p(t) + RP^p(t) \tag{5.2}$$

in which

$TP^p(t)$ the total net load of water pollutant p of the reservoir at time t (endogenous);
$WP_j^p(t)$ load of water pollutant p in region j at time t (endogenous);
$FP^p(t)$ load of water pollutant p by fisheries at time t (endogenous);
$SP_m^p(t)$ water pollutant p discharged by the sewage plant m (endogenous) and
$RP^p(t)$ water pollutant p of the rainfall that flows into the area (exogenous). Superscript p represents T-N if p = 1, T-P if p = 2, and COD if p = 3.

5.1.2.2 Water Pollutants Flow Through Rivers

All the water pollutant emitted by socio-economic activities flow through revivers.

$$WP_j^p(t) = \kappa_j \cdot SEP_j^p(t) \tag{5.3}$$

in which

κ_j current velocity (exogenous) and
$SEP_j^p(t)$ water pollutant emitted by socio-economic activities that flowing into reservoir through river j at time t (endogenous).

5.1.2.3 Water Pollutant Emitted by Socio-economic Activities

The total water pollutants emitted by socio-economic activities in the catchment area are composed of pollutants from households, non-point sources and point sources. Pollutants caused by fisheries are specified in Eq. (5.10), because they are directly loaded into the reservoir. Also, pollutants discharged through sewage are specified in Eq. (5.12). All the variables in this formula are specified in Eqs. (5.5)–(5.9).

$$SEP_j^p(t) = HP_j^p(t) + NP_j^p(t) + EFP_j^p(t) \tag{5.4}$$

in which

$HP_j^p(t)$ water pollutant p (excluding those through sewage) loaded by households in region j at time t (endogenous);

$NP_j^p(t)$ water pollutant p emitted by non-point sources in region j (endogenous)
and

$EFP_j^p(t)$ water pollutant p emitted by socio-economic activities (excluding those
by fisheries) in region j (endogenous).

5.1.2.4 Water Pollutants Caused by Household Wastewater in Each Municipality

Water pollutants caused by household wastewater in each municipality is associated with the population of each municipality.

$$HP_j^p(t) = EH^p \cdot P_j^N(t) \tag{5.5}$$

in which

EH^p emission coefficient of water pollutant p by household wastewater without
sewage plant (exogenous) and

$P_j^N(t)$ population that do not use sewage plant in region j and discharges pollutants into river (endogenous).

5.1.2.5 Load of Water Pollutants Through Non-point Sources

Total load of water pollutants through no-point sources is decided by the land area
and coefficient of water pollutant emitted through different land use.

$$NP_j^p(t) = EL^{pk} \cdot L_j^k(t) \tag{5.6}$$

in which

EL^{pk} coefficient of water pollutant p emitted through land use k (exogenous)
and

$L_j^k(t)$ area of land use k in region j that emitted pollutants into river (endogenous).

5.1.2.6 Load of Water Pollutants from Production Activities

Total load of water pollutants by production activities are composed by six industries pollutants cased by fisheries industry is specifies in Eq. (5.10). In order to
evaluate the treatment efficiency of two advanced technologies (*Biomass Recycle*

Plant & New Energy Project) and present technology, we set up the Eq. (5.9), especially to show the pollution emitted by pig farming industry.

$$EFP_j^p = EFP_j^{p3\sim6}(t) + EFP_j^{p2}(t) \tag{5.7}$$

$$EFP_j^{p3\sim6}(t) = EI^{pm} \cdot X_j^m(t) \quad (m \neq 1,2) \tag{5.8}$$

$$EFP_j^{p2}(t) = \sum_j EI^{pWF} \cdot NA_j^{WF}(t) + \sum_j EI^{pN} \cdot NA_j^N(t)$$
$$+ \sum_j EI^{pNP} \cdot NA_j^{NP}(t) + \sum_j EI^{BP} \cdot NA_j^{BP}(t) \tag{5.9}$$

in which

$EFP_j^{p3\sim6}(t)$ water pollutant p emitted by socio-economic activities (excluding pig farming and fisheries) in region j (endogenous);

$EFP_j^{p2}(t)$ water pollutant p emitted by pig farming industry in region j (endogenous);

EI^{pm} coefficient of water pollutant p emitted by industry m (exogenous);

$X_j^m(t)$ production of industry m in the area of region j that emitted pollutants into the rivers (endogenous);

EI^{pWF} coefficient of water pollutant p emitted in pig feces and urine treated by wet methane fermentation (exogenous);

$NA_j^{WF}(t)$ number of pigs for which feces and urine were treated by wet methane fermentation in region j (endogenous);

EI^{pN} coefficient of water pollutant p emitted in pig feces and urine without any treatment (exogenous);

$NA_j^N(t)$ number of pigs for which feces and urine were not treated in region j (endogenous);

EI^{pNP} coefficient of water pollutant p emitted by pigs whose feces and urine treated by *New Energy Project* (exogenous);

$NA_j^{NP}(t)$ number of pigs whose feces and urine treated by *New Energy Project* in region j (endogenous);

EI^{pBP} coefficient of water pollutant p emitted by pigs whose feces and urine treated by *Biomass Plant* (exogenous) and

$NA_j^{BP}(t)$ number of pigs whose feces and urine treated by *Biomass Plant* in region j (endogenous).

5.1.2.7 Load of Water Pollutants by Fisheries

Total load of water pollutants by fisheries is decided by the production and coefficient of water pollutant emitted by fisheries.

$$FP^p(t) = EF^{pf} \cdot X^f(t) \tag{5.10}$$

in which

EF^{pf} coefficient of water pollutant p emitted by fisheries (exogenous) and
$X^f(t)$ production of fisheries in the area (endogenous).

5.1.2.8 Load of Water Pollutant of the Rainfall

The total load of water pollutant of the rainfall is given as follows:

$$RP_j^p(t) = ER^p \cdot L^r \tag{5.11}$$

in which

ER^p coefficient of water pollutant p emitted by rainfall (exogenous) and
L^r the catchment's area of the area (endogenous).

5.1.2.9 Water Pollutant Treated by Sewage System

Water pollutant emitted by household is separated by sewage system and without any treatment, and the total load of water pollutant treated by sewage system is associated with the population that uses sewage system.

$$SP^p(t) = ES^p \cdot \sum_j P_j^{SP}(t) \tag{5.12}$$

in which

ES^p the pollution emission coefficient of sewage plant in the area at time t (exogenous) and
$P_j^{SP}(t)$ the population that use sewage plant of region j (exogenous).

5.1.3 Air Pollutants Flow Balance Model

The air pollutants flow balance model describes how the greenhouse gases are emitted in the catchment area. The air pollutants measured in this study included three greenhouse gases, CO_2, CH_4 and N_2O. CH_4 and N_2O have a much greater greenhouse effect potential than CO_2. Therefore, to appropriately evaluate emission of greenhouse gas, we have to consider the greenhouse effect potential of each gas.

The emission of greenhouse gases of each industry (except pig farming industry) is decided by the production and emission coefficient. When we introduce the two advanced technologies for pig farming industry, its emission of greenhouse gases is dependent on the number of pigs that treated by the advanced technologies, wet methane fermentation and no treatment. This formula has the relationship with Eqs. (5.9), (5.19) and (5.21)–(5.24).

$$TGG(t) = TG_1(t) + TG_4(t) \cdot GWP_4 + TG_5(t) \cdot GWP_5 \qquad (5.13)$$

$$TG_a(t) = EG_a^m \cdot \sum_j \{X_j^3 + X_j^4 + X_j^5 + X_j^6 + X_j^7\}$$

$$+ \{EG_a^{WF} \cdot \sum_j NA_j^{WF}(t) + EG_a^N \cdot \sum_j NA_j^N(t) + EG_a^{NP} \cdot \sum_j NA_j^{NP}(t)$$

$$+ EG_a^{BP} \cdot \sum_j NA_j^{BP}(t)\} + EG_a^C \cdot C(t)$$

$$(5.14)$$

in which

$TGG(t)$ total emission of greenhouse gas while considering the greenhouse effect potential of each gas (named TGG) at time t (endogenous) and

$TG_a(t)$ total emission of greenhouse gas or air pollutant a at term t (endogenous). Superscript a represents CO_2 if a = 1, CH_4 if a = 2, and N_2O if a = 3.

GWP_4 potential index of greenhouse effect of each gas when greenhouse effect by discharge of one molecule of carbon dioxide is assumed to be one (exogenous);

EG_a^m coefficient of greenhouse gas and air pollutant a emitted by industry m (m $\neq$ 2) (exogenous);

EG_a^{WF} coefficient of greenhouse gas and air pollutant a emitted by pig whose feces and urine treated by wet methane fermentation (exogenous);

EG_a^N coefficient of greenhouse gas and air pollutant a emitted by pigs whose feces and urine was not treated by any method (exogenous);

EG_a^{NP} coefficient of greenhouse gas and air pollutant a emitted by pigs whose feces and urine were treated by *New Energy Project* (exogenous);

EG_a^{BP} coefficient of greenhouse gas and air pollutant a emitted by pigs whose feces and urine were treated by *Biomass Plant* (exogenous);

EG_a^C coefficient of greenhouse gas and air pollutant a emitted by final demand (exogenous) and

$C(t)$ total consumption at time t (endogenous).

5.1.4 The Quantitative Change in Pig

In this paper, we calculated the quantitative change of pig as the relationship between mother pig and total amount pigs, in order to indicate the variation of total amount of pigs and the amount of pig farming wastes in the study area.

$$TMQ_j(t + 1) = TMQ_j(t) + \Delta TMQ_j(t) - d^p \cdot TMQ_j(t) \qquad (5.15)$$

in which

$TMQ_j(t)$ total amount of mother pig slaughtered of region j at time t (endogenous);

$\Delta TMQ_j(t)$ increase in the amount of mother pig slaughtered of region j at time t (endogenous) and

d^p depreciation rate of pig (exogenous).

The increase in mother pig is decided by the investment and reciprocal of the necessary investment per mother pig in the catchment area.

$$\Delta TMQ_j(t) = I_j^P(t) \cdot ITMQ \qquad (5.16)$$

in which

$I_j^P(t)$ the investment of mother pig of region j at time t (exogenous) and

$ITMQ$ reciprocal of the necessary investment per mother pig (exogenous).

The total amount of pig in the catchment area is equal to the sum of amount of pig slaughtered in each region.

$$TAQ(t) = \sum_j TAQ_j(t) \qquad (5.17)$$

in which

$TAQ(t)$ the total amount of pig of upstream area of Miyun reservoir of Miyun County at time t (endogenous).

The number of pig slaughtered is associated with the increase in the amount of pig slaughtered and the decrease in the amount of mother pig. The increase in the amount of pig slaughtered is specified in Eq. (5.20).

$$TAQ_j(t + 1) = TAQ_j(t) + \Delta TAQ_j(t) - d^{mp} \cdot d^p \cdot TMQ_j(t) \qquad (5.18)$$

in which

$TAQ_j(t)$ the total amount of pig slaughtered of region j at time t (endogenous);

$\Delta TAQ_j(t)$ increase in the amount of pig slaughtered of region j at time t (endogenous) and

d^{mp} the decrease amount of pig slaughtered when one mother pig decreases (exogenous).

The total amount of pig slaughtered is composed by the amount of pig whose feces and urine treated without any treatment, with wet methane fermentation technology, with *Biomass Recycle Plant* and with *New Energy Project*, respectively. The variables are specified in Eqs. (5.21)–(5.24).

$$TAQ_j(t) = NA_j^{WF}(t) + NA_j^{NP}(t) + NA_j^{BP} + NA_j^{N}(t) \qquad (5.19)$$

in which

$NA_j^N(t)$ number of pig whose feces and urine treated without any treatment method in region j (exdogenous).

The increase in amount of pig slaughtered is dependent on the amount of mother pig and farrowing rate of the mother pig.

$$\Delta TAQ_j(t) = TMQ_j(t) \cdot EA \tag{5.20}$$

in which

EA increase in amount of pig by per mother pig (exogenous).

$$NA_j^{NP}(t+1) = NA_j^{NP}(t) + \Delta NA_j^{NP}(t) \tag{5.21}$$

in which

$\Delta NA_j^{NP}(t)$ increase in amount of pig whose feces and urine treated by new energy plant in region j (endogenous).

$$NA_j^{BP}(t+1) = NA_j^{BP}(t) + \Delta NA_j^{BP}(t) \tag{5.22}$$

in which

$\Delta NA_j^{BP}(t)$ increase in amount of pig whose feces and urine treated by *Biomass Plant* in region j (endogenous).

The increase in amount of pig whose feces and urine treated by *New Energy Project* is dependent on the investment for the *New Energy Project*.

$$\Delta NA_j^{NP}(t) = I_j^{NP}(t) \cdot \Gamma^{NP} \tag{5.59}$$

in which

$I_j^{NP}(t)$ the investment for construction of *New Energy Project* to treat pig waste of region j at time t (exogenous) and
Γ^{NP} reciprocal of the necessary investment of *New Energy Project* per pig (exogenous).

The increase in amount of pig whose feces and urine treated by *Biomass Recycle Plant* is dependent on the investment for the *Biomass Recycle Plant*.

$$\Delta NA_j^{BP}(t) = I_j^{BP}(t) \cdot \Gamma^{BP} \tag{5.24}$$

in which

$I_j^{BP}(t)$ the investment for construction of *Biomass Plant* to treat pig waste of region j at time t (exogenous) and
Γ^{BP} reciprocal of the necessary investment of *Biomass Plant* per pig (exogenous).

The production of pig farming industry is associated with the amount of pig slaughtered.

$$X_j^2(t) \leq TAQ_j(t) \cdot \phi^p \tag{5.25}$$

in which

ϕ^p output coefficient of pig farming industry (exogenous).

5.1.5 Energy Balance Model

The energy balance model describes the amount of biomass energy that produced by two advanced technologies in the catchment area.

5.1.5.1 Energy Demand

Energy demand is decided by production of industrial section and consumption of final demand section.

$$ED_j^m(t) = \theta^m \cdot X_j^m(t) \tag{5.26}$$

$$ED^C(t) = \theta^C \cdot C(t) \tag{5.27}$$

in which

$ED_j^m(t)$ energy demand of industry m in region j at term t (endogenous);
θ^m the coefficient of energy demand of industry m (exogenous);
$ED^C(t)$ energy demand of final demand section at time t (exogenous) and
θ^C the coefficient of energy demand of final demand section (exogenous).

5.1.5.2 Total Energy Demand

Total energy demand consists of energy demand of industries and final demand section. The variables are specified in Eqs. (5.26) and (5.27).

$$TED(t) = \sum_j \sum_m ED_j^m(t) + ED^C(t) \tag{5.28}$$

in which

$TED(t)$ the total amount of energy demand at time t (endogenous).

5.1.5.3 Amount of Electric Energy Generated by Advanced Technologies

The amount of electric energy generated by two advanced technologies is dependent on the capital of two advanced technologies. The capitals of two advanced technologies are specified in Eqs. (5.30)–(5.35).

$$TEN_j(t) = \delta^{NP} \cdot K_j^{NP}(t) + \delta^{BP} \cdot K_j^{BP}(t) \tag{5.29}$$

in which

$TEN_j(t)$ amount of electric energy generated by *New Energy Project* in region j at time t (endogenous);

δ^{NP} the coefficient of electric energy generation of *New Energy Project* (exogenous);

$K_j^{NP}(t)$ the capital of *New Energy Project* available in region j at time t (endogenous);

δ^{BP} the coefficient of electric energy generation of *Biomass Plant* (exogenous) and

K_j^{BP} the capital of *Biomass Plant* available in region j at time t (endogenous).

5.1.5.4 Capital of Advanced Technologies

The capital of two advanced technologies is associated with the number of pigs that treated by the technologies. In this formula, the number of 10,000 and 1,000 represent the treatment scale of two technologies are 10,000 pigs' wastes and 1,000 pigs' wastes, respectively. The amount of pigs whose feces and urine treated by two advanced technologies are specifies in Eqs. (5.21) and (5.22).

$$K_j^{NP}(t) \geq \chi^{NP} \cdot \frac{NA_j^{NP}(t)}{10,000} \tag{5.30}$$

$$K_j^{BP}(t) \geq \chi^{BP} \cdot \frac{NA_j^{BP}(t)}{1,000} \tag{5.31}$$

in which

χ^{NP} the coefficient of *New Energy Project* cost per pig farm (exogenous) and
χ^{BP} the coefficient of *Biomass Plant* cost per pig farm (exogenous).

5.1.5.5 Capital Accumulation of Advanced Technologies

The capital accumulation of two advanced technologies is dependent on the investment for two advanced technologies. The construction investments for two advanced technologies are determined by construction allotment of the region and the subsidy that is provided by the government for the allotment.

$$K_j^{NP}(t+1) = K_j^{NP}(t) + I_j^{NP}(t) \tag{5.32}$$

$$I_j^{NP}(t) = \frac{1}{\gamma^{NP}} \cdot S_j^{NP}(t) \tag{5.33}$$

$$K_j^{BP}(t+1) = K_j^{BP}(t) + I_j^{BP}(t) \tag{5.34}$$

$$I_j^{BP}(t) = \frac{1}{\gamma^{BP}} \cdot S_j^{BP}(t) \tag{5.35}$$

in which

$I_j^{NP}(t)$ the investment for construction of *New Energy Project* in region j at time t (endogenous);

γ^{NP} rate of subsidization for construction of *New Energy Project* by the Miyun government (exogenous);

$S_j^{NP}(t)$ the subsidization for construction of *New Energy Project* by the Miyun government (endogenous);

$I_j^{BP}(t)$ the investment for construction of *Biomass Plant* in region j at time t (endogenous);

γ^{BP} rate of subsidization for construction of *Biomass Plant* by the Miyun government (exogenous) and

$S_j^{BP}(t)$ the subsidization for construction of *Biomass Plant* by the Miyun government (endogenous).

5.1.5.6 Energy Balance

The demand for energy of pig farming is covered by the amount of power generated by advanced technologies and the existing electric power industry.

$$TEN_j(t) + TED_j^2(t) \geq ED_j^2(t) \tag{5.36}$$

in which

$TED_j^2(t)$ the amount of electric energy of region j generated by electric power industry that provide to pig farming of Miyun county at time t (endogenous) and

$ED_j^2(t)$ the energy demand of pig farming in region j at time t (endogenous).

5.1.6 Treatment Policies for Non-point Sources

Miyun government developed the *Organic Fertilizer Promotion Policy* in 2004 and converted the majority of upland cropping into forest land to improve the water conservation function of Miyun Reservoir. This treatment policy is subsidized by the government for the promotion of organic fertilizers.

5.1.6.1 Total Land Area

Total land area is separated by five land use forms in each region.

$$\overline{L}_j(t) = \sum_{k=1}^{5} L_j^k(t) \tag{5.37}$$

in which

$\overline{L}_j(t)$ total land area in region j at time t (endogenous) and

$L_j^k(t)$ different types of land use area in region j at time t (endogenous).

5.1.6.2 Conservation of Other Land to Forest Land

Miyun government converted the majority of upland cropping into forest land to improve the water conservation function of Miyun Reservoir.

$$L_j^k(t + 1) = L_j^k(t) + \Delta L_j^k(t) \tag{5.38}$$

in which

$\Delta L_j^k(t)$ increase in forest land that conversed from other land use in region j at time t (endogenous) and $k = 2$

$$\Delta L_j^k(t) = L_j^{52}(t) \tag{5.39}$$

in which

$L_j^{52}(t)$ conversion of land use from other land (superscript $= 5$) to forest land (superscript $= 2$) in region j at time t (endogenous) and $k = 2$.

The conversion from other land area to forest land is subsidized by the Miyun government.

$$L_j^{52}(t) \geq \lambda^5 \cdot S_j^{5f}(t) \tag{5.40}$$

in which

$L_j^{52}(t)$ conversion of land use from other land area to forest land in region j at time t (endogenous);

λ^5 reciprocal of the subsidy for one unit conversion to forest (exogenous) and

$S_j^{5f}(t)$ subsidy given by the government for conversion of land use (endogenous).

The change rate of variable $L_j^k(t)$ was indicated as endogenous variable $\Delta L_j^k(t)$ to indicate how the variable changes from year t to t + 1. The change rate was constrained and depended on the value of endogenous variable $S_j^{5f}(t)$ which can

be achieved through automatic calculation process (looping computation and iterations) based on the interactive and restrictive correlations among variables.

5.1.6.3 Conservation of Land Area with Organic Fertilizer

Miyun government developed the *Organic Fertilizer Promotion Policy* in 2004.
This treatment policy is subsidized by the government for the promotion of
organic fertilizers.

$$L_j^1(t) = LC_j^{IF}(t) + LC_j^{OF}(t) \tag{5.41}$$

in which

$L_j^1(t)$ the total area of planting in region j at time t (endogenous);

$LC_j^{IF}(t)$ the area of planting which applied with inorganic fertilizer in region j at
time t (endogenous) and

$LC_j^{OF}(t)$ the area of planting which applied with organic fertilizer in region j at
time t (endogenous).

$$LC_j^{OF}(t+1) = LC_j^{OF}(t) + \Delta LC_j^{OF}(t) \tag{5.42}$$

in which

$\Delta LC_j^{OF}(t)$ increase in planting area which applied with organic fertilizer in
region j at time t (endogenous).

   The increase in planting area which applied with organic fertilizer is decided by
the subsidization provided by the Miyun government.

$$\Delta LC_j^{OF}(t) = ILC_j^{OF} \cdot S_j^{OF}(t) \tag{5.43}$$

in which

ILC_j^{OF} increasing amount of planting area which applied with organic fertilizer
per thousand RMB Yuan in region j at time t (exogenous) and

$S_j^{OF}(t)$ subsidization for the utilization of organic fertilizer in region j at time t.
(endogenous).

5.1.7 Policies for Household Wastewater Generation

According to the characteristics and situations of suburbs of big cities, the treatment of household wastewater mainly depends on the sewage systems and sewage
plants. These measures are implemented by the local governments.

5.1.7.1 Change in Population

The change in population in each municipality is given as follows:

$$P_i(t+1) = P_i(t) + \Delta P_i(t) \tag{5.44}$$

in which

$P_i(t)$ population of municipality i at time t (endogenous) and
$\Delta P_i(t)$ increase in the population of municipality i at time t (endogenous).

The population can be separated as that use sewage plant and without any wastewater treatment. According to this formula, the Eqs. (5.5) and (5.12) can be implemented.

$$P_i(t) = P_i^{SP}(t) + P_i^{N}(t) \tag{5.45}$$

in which

$P_i^{SP}(t)$ the population that use sewage plant of municipality i (endogenous) and
$P_i^{N}(t)$ the population that without any wastewater treatment of municipality i (endogenous).

$$\Delta P_i(t) = \Delta P_i^{SP}(t) + \Delta P_i^{N}(t) \tag{5.46}$$

in which

$\Delta P_i^{SP}(t)$ increase in the population that use the sewage plant (endogenous) and
$\Delta P_i^{N}(t)$ increase in the population that without any wastewater treatment (endogenous).

5.1.7.2 Increase in the Population that Uses the Sewage System and Plant

The increase in the population that used the sewage system and plant are dependent on the construction investment which is specified by Eqs. (5.49)–(5.52).

$$\Delta P_i^{SP}(t) \leq \lambda_i^{SP} \cdot I_i^{SP}(t) \tag{5.47}$$

in which

λ_i^{SP} reciprocal of the necessary construction investment per person that uses the sewage plant (exogenous) and
$I_i^{SP}(t)$ construction investment of municipality i for sewage plant (endogenous).

5.1.7.3 Local Finance

Majority of construction investment and maintenance cost of sewage plant are dependent on the local finance of each municipality.

$$FE_i(t+1) = FE_i(t) \cdot (1 + \theta_i) \tag{5.48}$$

in which

$FE_i(t)$ total finance of municipality i (endogenous) and
θ_i the rate of total finance growth of municipality i (exogenous).

5.1.7.4 Sewage System

Investments for construction of sewage systems and plants are determined by the construction allotment of the municipality and subsidies that are provided by the local government. The construction allotment and maintenance costs are covered by local finances and special subsidizations.

$$K_i^{SP}(t+1) = K_i^{SP}(t) + I_i^{SP}(t) \tag{5.49}$$

$$I_i^{SP}(t) = (\frac{1}{1 - M_i^{SP}}) \cdot \xi_i^{SP} \cdot FE_i(t) \tag{5.50}$$

in which

$K_i^{SP}(t)$ capital available for sewage system in municipality i at time t (endogenous);
M_i^{SP} rate for investment of sewage system from the municipality i (exogenous) and
ξ_i^{SP} rate transfer for investment of household wastewater from the total finance of municipality i (exogenous).

The maintenance cost of sewage system is mainly provided by local finance. The total investment and maintenance cost for construction of sewage system are dependent on the local finance and household wastewater treatment subsidy that granted by the Miyun government to the municipality for intensive promotion of construction and installation for treatment measures of household wastewater. According to the current situation in China, there is no local bond and majority of the maintenance of sewage system is derived from local finance.

$$MC_i^{SP}(t) = \varsigma_i^{SP} \cdot FE_i(t) \tag{5.51}$$

$$I_i^{SP}(t) + MC_i^{SP}(t) \leq \varpi \cdot FE_i(t) + S_i^{SP}(t) \tag{5.52}$$

in which

MC_i^{SP} maintenance of sewage system of municipality i at time t (endogenous);
ς_i^{SP} rate transfer for maintenance cost of household wastewater from the total finance of municipality i (exogenous);
ϖ rate transfer for treatment of household wastewater from the total finance of municipality i (exogenous) and

$S_i^{SP}(t)$ subsidy for households wastewater treatment in municipality i, that is granted by Miyun government (endogenous).

5.1.8 Treatment Measures for Production Generation Sources

5.1.8.1 Production Function and Curtailment

This production function is derived from Harrod-Domar model through the relationship between capital accumulation and production. We assumed the production of industry m is restricted by leaving capital idle and subsidy for loss due to the idle capital.

$$X_j^m(t) \leq \alpha^m \cdot \{K_j^m(t) - S_j^m(t)\} \tag{5.53}$$

in which

α^m ratio of capital to output in industry m (exogenous).

The capital accumulation is dependent on the investment and depreciation of capital.

$$K_j^{mP}(t+1) = K_j^{mP}(t) + I_j^{mP}(t+1) - d^m \cdot K_j^{mP}(t) \tag{5.54}$$

in which

$K_j^{mP}(t)$ capital available for industry m in region j at time t (endogenous);
$I_j^{mP}(t)$ investment in industry m in region j at time t (endogenous) and
d^m depreciation rate of industry m (exogenous).

5.1.9 Total Budget of the Government for the Countermeasures

It is assumed that the Miyun government spends 149.3 million RMB Yuan for implementing the countermeasures every year. This figure is based on the actual budget that has been directly and indirectly spent to improve the quality of the reservoir. These variables are specified in Eqs. (5.52), (5.53), (5.43), (5.40), (5.33) and (5.35).

$$\begin{aligned} y(t) \geq \sum_i S_i^{SP}(t) + \sum_j \sum_m S_j^m(t) + \sum_j S_j^{OF}(t) \\ + \sum_j S_j^{5f}(t) + \sum_j S_j^{NP}(t) + \sum_j S_j^{BP}(t) \end{aligned} \tag{5.55}$$

in which

$y(t)$ the total budget spent by the local government for implementing the countermeasures (exogenous).

5.1.10 Flow Balance in the Commodity Market

Each industry must produce in order to meet the balance between supply and demand for the commodity produced in the industry. The production is dependent on Leontief input-output coefficient matrix, consumption, investment and net export. In this equation, we added the variables related to the investment of two advanced technologies, in order to describe the influence of the introduction of two advanced technologies on the production, when we introduce the two advanced technologies. The total consumption is specified in Eq. (5.58), investment for industry, sewage plant, *New Energy Project* and *Biomass Recycle Plant* are specified in Eqs. (5.54), (5.50), (5.32) and (5.34), the capital of two advanced technologies are specifies in Eqs. (5.30) and (5.31), the net export represented as Eq. (5.57), respectively.

$$X(t) \geq A \cdot X(t) + C(t) + i^m(t) + B^{SP} \cdot I^{SP} + B^{NP} \cdot I^{NP}(t) + B^{BP} \cdot I^{BP}(t)$$
$$+ B^{mNP} \cdot \sum_j K_j^{NP}(t) + B^{mBP} \cdot \sum_j K_j^{BP}(t) + e(t) \tag{5.56}$$

in which

$X(t) = \sum_j X_j(t)$	column vector of the m-th element that is the total product of industry m in the basin (endogenous);
A	input-output coefficient matrix (exogenous);
$C(t)$	total consumption at time t (endogenous);
$i^m(t) = \sum_j I_j^{mP}(t)$	total investment at time t (exogenous);
B^{SP}	column vector of m-th coefficient that induced production in industry m by construction of sewage plant (exogenous);
B^{NP}	column vector of m-th coefficient that induced production in industry m by construction of *New Energy Project* (exogenous);
B^{BP}	column vector of m-th coefficient that induced production in industry m by construction of *Biomass Plant* (exogenous);
B^{mNP}	column vector of m-th coefficient that induced production in industry m by maintenance of *New Energy Project* (exogenous);
B^{mBP}	column vector of m-th coefficient that induced production in industry m by maintenance of *Biomass Plant* (exogenous) and
$e(t)$	column vector of net export (endogenous).

5.1.11 Restriction of Net Export

The net export is restricted as follows:

$$e_{\min} \leq e(t) \leq e_{\max} \tag{5.57}$$

in which

$e_{min}(t)$ column vector of minimum net export (exogenous) and
$e_{max}(t)$ column vector of maximum net export (exogenous).

5.1.12 Restriction of Consumption

The restriction of consumption is dependent on the population and column vector of the consumption of each industry.

$$C(t) \geq H \cdot \sum_i P_i(t) \tag{5.58}$$

in which

H column vector of the m-th element that is the consumption of industry m in the basin (exogenous).

5.1.13 Gross Regional Product

We adopt the index of GRP to reflect the development of the local socio economy, which is dependent on the production and added value of each industry.

$$GRP(t) = \upsilon \cdot X(t) \tag{5.59}$$

in which

$GRP(t)$ gross regional product (endogenous) and
υ row vector of m-th element that is rate of added value in m-th industry (exogenous).

5.1.14 Constraints on the Regional Economy

It is assumed that the GRP realize annual increase or at least equals the GRP of last year. This formula is based on the actual characteristics and requirements of economic development in China.

$$GRP(t+1) \geq GRP(t) \tag{5.60}$$

5.1.15 Targeted Environmental Improvement

There is one restriction on the emission of TGG for each year.

$$TGG(t) \leq \overline{TGG(t)} \tag{5.61}$$

in which

$\overline{TGG(t)}$ the emission restriction of TGG at time t (exogenous).

5.2 Simulation Results

In this study, we defined the simulation cases that reflect economic development level (GRP) as increases of n% in 2016 compared to 2006. Representative cases were selected to indicate the variations in the simulation results. Case 10 represents a GRP in 2016 with an increase of 10 % compared to 2006; Case 15 an increase of 15 % compared to 2006; Case 20 an increase of 20 % compared to 2006.

5.2.1 Objective Function and Total GRP

In this study, we adopted integrated policies with the introduction of two advanced technologies for the simulation, and the policy effectively reduced the environmental pollutants in this simulation. A feasible solution was reached for Case 10 that both adopted and did not adopt the two advanced technologies, and a feasible solution was achieved for Case 20 when we adopted the two advanced technologies. No feasible solution could be obtained for Case 21 that both adopted and did not adopt the two advanced technologies.

Comparing the results of the simulations, the introduction of advanced technologies is effective to reduce water pollutants and air pollutants, and at the same time achieve economic development. When introduction of the advanced technologies and GRP increased by 20 % compared to 2006 (Case 20), the objective value for T-N (sum of T-N for 10 years) decreased about 3,499 t as compared to when not adopted (Fig. 5.2). Introduction of two advanced technologies raised the rate of economic growth 10 % as compared to no introduction of advanced technologies. Moreover, the total GRP accumulated for 10 years had a difference of more than 1,819 million RMB Yuan when the

Fig. 5.2 Objective value for T-N (sum of T-N for 10 years)

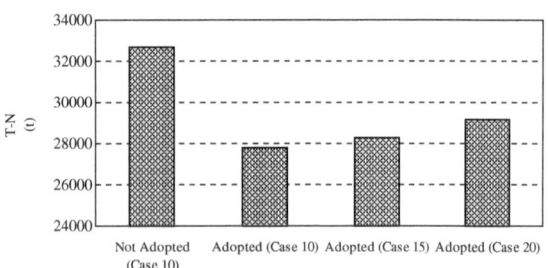

Fig. 5.3 Accumulated total
GRP (2007–2016)

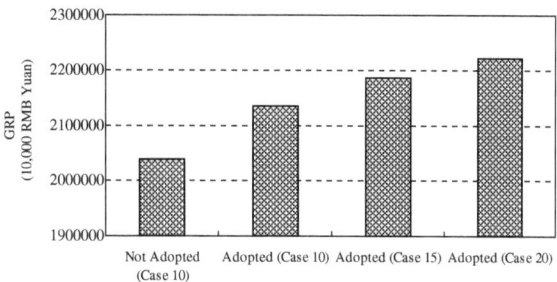

Fig. 5.4 Changes in T-N

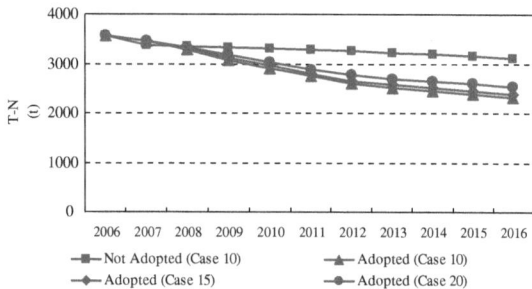

two advanced technologies were adopted in Case 10 as compared to when not adopted (Fig. 5.3).

The variations in total amounts of T-P and COD flowing into the Miyun Reservoir are shown in Figs. 5.5 and 5.6. In Case 10, the net load of T-N was reduced 25.3 % (see Fig. 5.4), and the net load of T-P and COD was reduced 42.0 and 36.7 %, respectively, in 2016 with adoption of the advanced technologies when compared with no adoption.

This result verifies that it is necessary to consider the minimization of T-N as the objective function when we formulate integrated policies to improve the water environment.

5.2.2 Budget Expenditures for the Policy

In Fig. 5.7, we show the accumulation of budget allotment in the simulation. It is interesting that the measured cost for the production generation source decreased when the advanced technologies were adopted in Case 10–20 as compared to when not adopted (Case 10) and subsidies were drawn from the introduction of the two advanced technologies. This result shows that the subsidy for the introduction of two advanced technologies can mainly be covered by the diverted budget from the measured costs for the production generation source.

Fig. 5.5 Changes in T-P

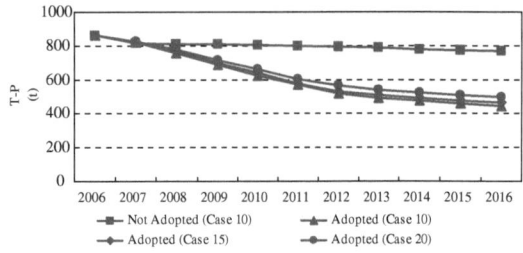

Fig. 5.6 Changes in COD

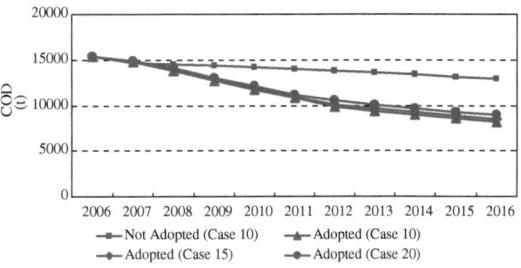

As shown in Fig. 5.7, we also found that the measured costs for household wastewater generation have only a small difference among Cases 10–20 that adopted the advanced technologies. Therefore, we conclude that a policy that installs more sewage plants and systems is an effective tool to reduce pollutants emitted by households and this policy is adaptive to specific situation in rural areas of China.

From the aspects of different regions, installation of advanced technologies was performed more in Region 2 that flows into the Chao River. Based on the situation in 2006, the towns of Region 2 emitted 2.7 times more T-N into the catchment area of Miyun Reservoir than Region 1 in the basin and the population of Region 2 is larger than Region 1, for example Gao Ling Town and Tai Shitun Town. Therefore, Region 2 should take more responsibility for the treatment of environmental pollution. However, its per capita scale of financial affairs is small. Even in this simulation, it is difficult to introduce advanced technologies by the self-financial affairs because the financial scale is small and most of the budget from the Miyun government was provided to Region 2. Therefore, the Miyun government should pay more attention and provide a larger budget for the installation of sewage systems and advanced technologies in Region 2.

5.2.3 Variations in Production of Pig Farming Industry

The variation in production of the pig farming industry is shown in Fig. 5.8. In this simulation, the policy to introduce advanced technologies is adopted in Case 20 in

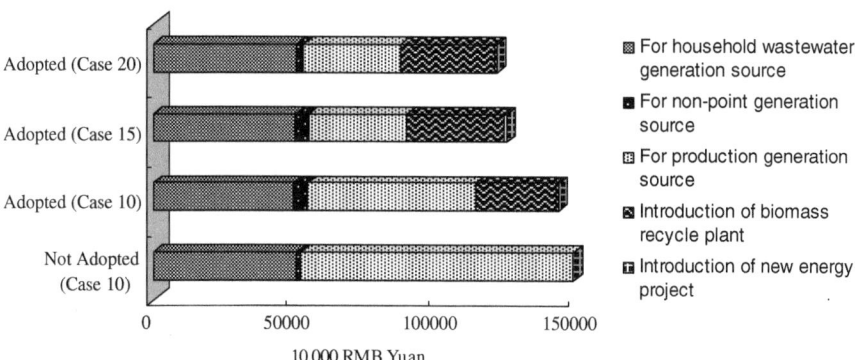

Fig. 5.7 Accumulated total budget expenditures for each policy

Fig. 5.8 Changes in
production of pig farming
industry

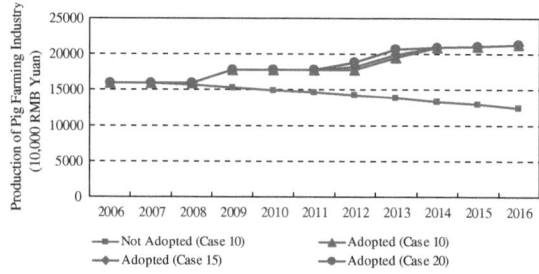

the catchment area, the amount of production of the pig farming industry increases
about 448 million RMB Yuan over a total of 10 years for the simulation period as
compared with not adopting the technologies and the production of pig farming
increased more than 33.7 % as compared to 2006 (Yan et al. 2010).

5.2.4 Pollutants Emitted from Pig Farming Industry

Table 5.1 shows that the water pollutants from the pig farming industry that
flow into the reservoir were obviously reduced with the introduction of the two
advanced technologies, 6,423 t of T-N, 2,716 t of T-P and 42,729 t of COD can
be reduced in the total 10 years (2007–2016) by introducing the advanced
technologies.

Moreover, from the aspect of greenhouse gases, the advanced technologies
emit much less CH_4 and N_2O than any other method of animal waste treatment.
This indicates that utilization of biomass resources contributes to mitigation of the
GHG, even if CO_2, a generated biomass resource is not considered as carbon neutral. In this simulation, about 17,000 t of greenhouse gas can be reduced in the
total 10 years by introducing the two advanced technologies (see Table 5.2).

Table 5.1 Reduced emissions of water pollutants with advanced technologies

		Water pollutants		
		T-N	T-P	COD
Reduced in emissions of water pollutants in pig farming industry = Derived emission amount in Case 10 without advanced technologies—Case 20 with advanced technologies	2006	0	0	0
	2007	53.1	24.1	382.7
	2008	175.7	75.9	1,195.8
	2009	308.8	131.7	2,073.1
	2010	441.8	187.5	2,950.4
	2011	574.8	243.3	3,827.8
	2012	707.8	299.1	4,705.1
	2013	840.9	355.0	5,582.5
	2014	973.9	410.8	6,459.8
	2015	1,106.9	466.6	7,337.2
	2016	1,239.9	522.4	8,214.5
Accumulated total in 10 Years (2007–2016)		6,423.6	2,716.4	42,729.0

Unit: t

Table 5.2 Reduced emissions of greenhouse gases with advanced technologies

		Green house gases			
		CO_2	CH_4	N_2O	GHG
Reduced in emissions of water pollutants in pig farming industry = Derived emission amount in Case 10 without advanced technologies—Case 20 with advanced technologies	2006	0	0	0	0
	2007	−433	96	1	2,130
	2008	−2,632	295	2	5,277
	2009	−5,094	510	3	8,577
	2010	−7,557	725	4	11,877
	2011	−10,019	939	6	15,176
	2012	−12,481	1,154	7	18,476
	2013	−14,943	1,369	8	21,775
	2014	−17,406	1,584	10	25,075
	2015	−19,868	1,799	11	28,375
	2016	−22,330	2,014	12	31,674
Accumulated total in 10 Years (2007–2016)		−112,764	10,485	64	168,413

Unit: t

5.2.5 Amount of Electric Energy Produced by the Advanced Technologies

The amount of electric energy produced by the two advanced technologies is shown in Table 5.3. In this simulation, we found that if a policy to introduce the advanced technologies is adopted in Case 20 in the catchment area, about 33 million kWh electrical energy over the total 10 years can be generated by two advanced technologies.

Table 5.3   Amount of electric energy produced by the two advanced technologies

Years	Electrical energy		
	Adopted (Case 10)	Adopted (Case 15)	Adopted (Case 20)
2007	37.94	33.63	28.66
2008	99.95	97.83	90.97
2009	161.96	162.02	158.30
2010	223.97	226.22	225.64
2011	285.98	290.41	292.97
2012	347.99	354.61	360.31
2013	410.00	418.81	427.64
2014	472.01	483.00	494.97
2015	534.02	547.20	562.31
2016	596.03	611.40	629.64
Accumulated total in 10 Years (2007–2016)	3,169.86	3,225.13	3,271.42

Unit: 10,000 kWh

5.3   Conclusions and Discussions

In this study, when we adopted the policy to introduce two advanced technologies (*New Energy Project* and *Biomass Energy Recycle Plant*), the policy was a very effective tool to reduce environmental pollutants in all simulations. The introduction of two advanced technologies raised the level of economic growth by 10 % as compared to not adopting the advanced technologies policy. When the two advanced technologies were introduced in Case 10, the objective value (total T-N) showed a reduction of about 4,916 t as compared to not adopting the advanced technologies. Moreover, we also found that if the policy to introduce the advanced technologies was adopted in the catchment area, production of the pig farming industry increased about 448 million RMB Yuan over the total ten year simulation period as compared with not adopting.

Concerning environmental pollutants, we found a reduction of about 168,000 t of greenhouse gases emitted by the pig farming industry in the total ten year simulation (2007–2016) when we introduced the advanced technologies in Case 20. In addition, we found that about 6,424 t of T-N, 2,716 t of T-P and 42,729 t of COD emitted by the pig farming industry in the total ten year simulation can be reduced by introducing the advanced technologies. From the aspect of biomass electric energy, more than 32.7 million kWh of electric energy can be generated when we adopted the two advanced technologies.

The results of this study establish the synthetic policies for the catchment area, especially the introduction of advanced technologies for pig farming industry are very effective and the utilization of biomass resources allows simultaneous pursuit of environmental preservation and economic development.

References

Yan JJ, Xu F, Kang CJ, Higano Y (2010) Effective stockbreeding biomass resource use and its impact on water environment from the viewpoint of sustainable development. J Dev Sustain Agric 5:147–150

Yan JJ, Sha JH, Chu X, Xu F, Higano Y (2014) Endogenous derivation of optimal environmental policies for proper treatment of stockbreeding wastes in the upstream region of the Miyun Reservoir, Beijing. Pap Reg Sci 93:477–500

Chapter 6
Cost Benefit Analysis and Regional Analysis

Abstract In this Chapter, we carried out the cost-benefit analysis and regional analysis to research how to arrange the two advanced technologies in the catchment area for different towns. The results of socio cost-benefit analysis show that there is great feasibility and possibility to introduce advanced technologies in the catchment area. From the aspects of regions, the application of advanced technologies is based on the concentration and scale of pig-farms in different regions.

Keywords Cost benefit analysis · Regional analysis · Application

6.1 Cost-Benefit Analysis

6.1.1 Socio Cost-Benefit Analysis

The simulation carried out in Case 20 shows that 151 *Biomass Recycle Plants* and 14 *New Energy Project*s should be installed to treat the pig wastes (see Fig. 6.1 and Table 6.1). The total cost is about 1.09 billion RMB Yuan over the total ten years. At the same time, the accumulated total GRP would increase about 1.82 billion RMB Yuan over the same period.

Beside this, we calculated the added value of advanced technologies as 4 parts, including biomass electric value, organic fertilizer value, CO_2 mitigation value and reduction of environmental fines. The detailed calculation was shown as follows (The price of CO_2 is based on the price in Chicago Climate Exchange):

© The Author(s) 2015
J. Yan, *Comprehensive Evaluation of Effective Biomass Resource Utilization and Optimal Environmental Policies*, SpringerBriefs in Economics, DOI 10.1007/978-3-662-44454-2_6

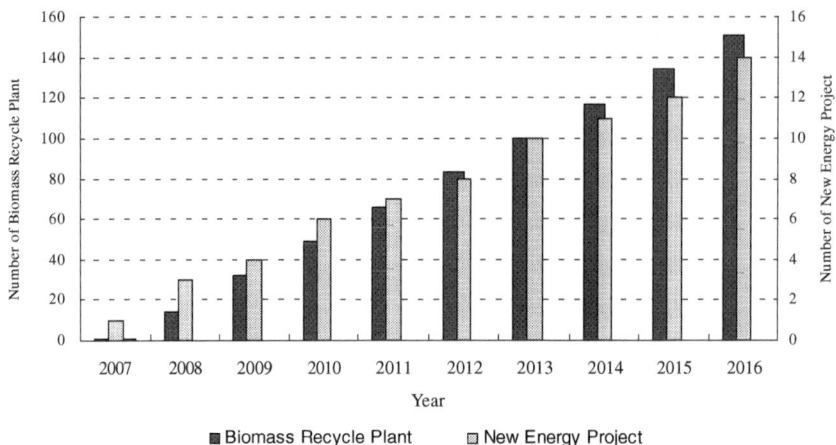

Fig. 6.1 Number of two advanced technologies (case 20)

Table 6.1 Regional allocation of advanced technologies in 2016

Region	Town	Number of Pig	Subsidy (1,000 RMB Yuan)	Per capita scale of financial affairs (1,000 RMB Yuan)	Number of new energy project	Number of *Biomass Plant*
Bai river	Shi Cheng	3,445	273.6	3.78	0	0
	Feng Jiayu	12,663	1,005.5	11.62	2	0
	Bu Laotun	60,459	4,800.7	1.11	6	0
Sub total		76,568	6,079.8	4.04	8	0
Chao river	Gao Ling	66,117	1,492.1	2.77	2	47
	Gu Beikou	19,345	436.6	12.18	0	14
	Xin Chengzi	22,197	500.9	2.47	1	16
	Tai Shitun	45,594	1,028.9	1.95	1	33
	Bei Zhuang	25,778	581.7	3.60	1	18
	Da Chengzi	32,077	723.9	1.61	1	23
Sub total		211,107	4,764.2	3.20	6	151
Total		287,675	10,844	3.62	14	151

The added value of advanced technologies

$$= \text{Biomass electric value} + \text{Organic fertilizer value} + CO_2 \text{ mitigation value}$$
$$+ \text{Reduction of environmental fines}$$
$$= 0.5 \text{ RMB Yuan/kwh} \times 33,000,000 \text{ kwh}$$
$$+ 100 \text{ RMB Yuan/t} \times 97,500 \text{ t}$$
$$+ 70 \text{ RMB Yuan/t} \times (112,764 + 33,000) \text{ t}$$
$$+ 912,000 \text{ RMB/year} \times 10 \text{ years}$$
$$= 16,500,000 + 9,750,000 + 10,203,480 + 9,120,000$$
$$= 45.57 \text{ million RMB Yuan}$$

Therefore, these results of socio cost-benefit analysis show that there is great feasibility and possibility to introduce advanced technologies in the catchment area.

6.1.2 *Private Cost-Benefit Analysis*

Private cost-benefit analysis is an important content and basis of the pig farms or enterprises to make the investment decision and evaluate the risk for introduction of advanced technologies, and is a main condition whether bank and other financial sector provide loan. We implemented the cost-benefit analysis to evaluate investment cost and benefit from private perspective for two advanced technologies. In this analysis, we evaluated three primary figures of merit for one unit technology: autonomous construction investment, depreciation and net benefit. The superscript represent the two advanced technologies (NP represents *New Energy Project*, BP represents *Biomass Recycle Plant*).

6.1.2.1 Autonomous Construction Investment

Autonomous investment is motivated by change in technology, resources and governmental policies and is independent of margin of profit. Here, we use autonomous construction investment to represent the amount of investment that provided by pig farm or enterprise for the introduction of advanced technologies after the subsidization provided by Miyun government. According to the environmental policies of Miyun government, it will provide 30 % of investment construction cost as subsidization for the introduction of high efficiency wastes treatment technology.

$$\text{Autonomous construction investment}^{NP}$$
$$= \text{construction investment cost}^{NP} \times (1 - \text{subsidization rate})$$
$$= 2.81 \times 70\,\%$$
$$= 1.97 \text{ million RMB Yuan}$$

$$\text{Autonomous construction investment}^{BP}$$
$$= \text{construction investment cost}^{BP} \times (1 - \text{subsidization rate})$$
$$= 6.5 \times 70\,\%$$
$$= 4.55 \text{ million RMB Yuan}$$

6.1.2.2 Maintenance Cost

The maintenance cost of *New Energy Project* is about 0.22 million RMB Yuan per year, and the maintenance cost of *Biomass Recycle Plant* is about 0.39 million RMB Yuan per year.

6.1.2.3 Annual Depreciation

We assumed that the salvage value percent is 10 %. This estimation is based on the official data published in 2002 by State Environmental Protection Administration. The service life of two technologies is 20 years.

$$\text{Annual Deprecation}^{NP}$$
$$= \left(\text{construction investment cos t}^{NP} - \text{subsidization}^{NP}\right)$$
$$\times \left(1 - \text{salvage value percent}^{NP}\right) \times \left(1/\text{service life}^{NP}\right)$$
$$= 1.97 \times 90\% \times (1/20)$$
$$= 0.089 \text{ million RMB Yuan}$$

$$\text{Annual Deprecation}^{BP}$$
$$= \left(\text{construction investment cost}^{BP} - \text{subsidization}^{BP}\right)$$
$$\times \left(1 - \text{salvage value percent}^{BP}\right) \times \left(1/\text{service life}^{BP}\right)$$
$$= 4.55 \times 90\% \times (1/20)$$
$$= 0.205 \text{ million RMB Yuan}$$

6.1.2.4 Direct and Indirect Technology Income

We calculated the direct and indirect technology income from three aspects, including biomass electric value, organic fertilizer value and reduction of environmental fines.

$$\text{Annual Income}^{NP}$$
$$= \text{biomass electric value}$$
$$+ \text{organic fertilizer value}$$
$$+ \text{reduction of environmental fines}$$
$$= 0.655 \text{ million RMB Yuan}$$

$$\text{Annual Income}^{BP}$$
$$= \text{biomass electric value}$$
$$+ \text{organic fertilizer value}$$
$$+ \text{reduction of environmental fines}$$
$$= 0.687 \text{ million RMB Yuan}$$

6.1.2.5 Net Benefit

Net benefit represents the total amount of net benefit of each technology during its service life.

Net benefit^{NP}

$$= \left(\text{Annual income}^{NP} - \text{Annual maintenance cost}^{NP} - \text{Annual depreciation}^{NP} \right)$$
$$\times \text{ service life}^{NP} - \text{autonomous construction investment}^{NP} + \text{salvage value}^{NP}$$
$$= (0.655 - 0.22 - 0.089) \times 20 - 1.97 + 0.281$$
$$= 6.231 \text{ million RMB Yuan}$$

Net benefit^{BP}

$$= \left(\text{Annual income}^{BP} - \text{Annual maintenance cost}^{BP} - \text{Annual depreciation}^{BP} \right)$$
$$\times \text{ service life}^{BP} - \text{autonomous construction investment}^{BP} + \text{salvage value}^{BP}$$
$$= (0.687 - 0.39 - 0.205) \times 20 - 4.55 + 0.65$$
$$= -2.06 \text{ million RMB Yuan}$$

The private cost-benefit analysis results represent that *New Energy Project* is more feasible and acceptable as compared to *Biomass Recycle Plant* from the aspect of private financial risk. However, because the data of maintenance cost of *Biomass Recycle Plant* is estimated based on that in Japan, when we introduce this technology in China, the maintenance cost and construction investment cost would be modified and reduced as the actual situation and labor price in the study area. *Biomass Recycle Plant* would become more feasible and acceptable.

Moreover, when the subsidization rate for the technology of *Biomass Plant* was increased to 47 %, the pig farm or enterprise that introduces *Biomass Plant* technology would be able to break even and start making profits.

6.2 Regional Analysis

Table 8 shows the regional allocation of advanced technologies for pig farming industry in 2016. We find another interesting phenomenon that the number of pigs in Region 2 is about 3 times larger than that in Region 1. But the subsidy provided for Region 1 is larger than Region 2. The subsidy provided for Region 1 mainly is occupied by Bu Laotun Town, and 6 *New Energy Projects* should be installed for the concentrated large-scale pig farms in Bu Laotun Town. Even in this simulation, since its per capita scale of financial affairs is the smallest one in this catchment area, installation of sewage system was not set up in time to keep up with the regional development. Therefore, the Miyun government should provide more budgets to introduce advanced technologies in Bu Laotun Town prior to other towns.

Figure 6.2 represents the key subsidy regions that include Bu Laotun, Gao Ling and Tai Shitun towns. In Bu Laotun Town, 6 *new energy projects* should be installed for concentrated large pig farms, so we call it A type town. In Gao Ling Town, 2 *New Energy Project* and 47 *Biomass Recycle Plants* should be installed

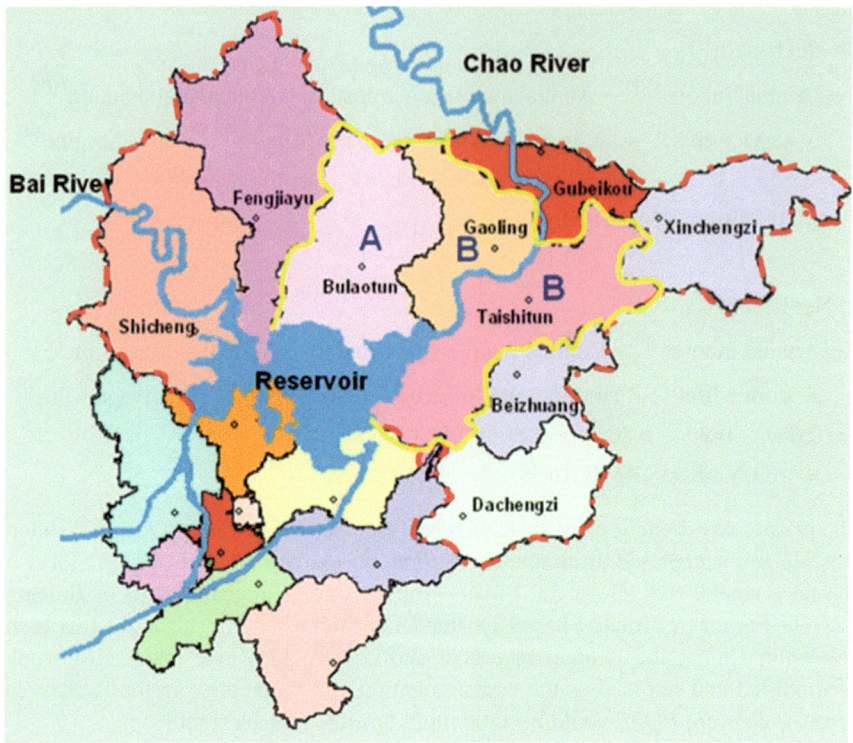

Fig. 6.2 Distribution of key subsidy regions (case 20)

for dispersed and small pig farms, therefore we name if B type town. In Tai Shitun Town, 1 *New Energy Project* and 33 *Biomass Recycle Plants* should be installed to satisfy the requirements of small pig farms, so it is also called B type town.

6.3 Conclusions and Discussions

In this Chapter, we carried out the cost-benefit analysis and regional analysis to research how to arrange the two advanced technologies in the catchment area for different towns.

The results of socio cost-benefit analysis show that there is great feasibility and possibility to introduce advanced technologies in the catchment area. The private cost-benefit analysis results represent that *New Energy Project* is more feasible and acceptable as compared to *Biomass Recycle Plant* from the aspect of private financial risk. And when the subsidization rate for the technology of *Biomass Plant* was increased to 47 %, the pig farm or enterprise that introduces *Biomass Plant* technology would be able to break even and start making profits.

From the aspects of different regions, although Chao River basin has smaller amount of pig, its large-scale pig farms are concentrated, more *New Energy Project* should be installed in Region 1; since the distribution of pig farms is dispersed in Bai River basin, more *Biomass Plant* should be installed in Region 2.

The results of this study establish the synthetic policies for the catchment area, especially the introduction of advanced technologies for pig farming industry are very effective and the utilization of biomass resources allows simultaneous pursuit of environmental preservation and economic development from the socio aspect. When we introduce advanced technologies to utilize biomass resource, preferential policies becomes essential, the government should implement preferential policies, such as carrying out tax preference and raising subsidized rate to encourage the utilization of advanced technologies. This comprehensive evaluation is expected to improve the biomass resource utilization and environmental reservation and form the basis of policy decision-making for sustainable development of rural areas of big cities in China (Yan et al. 2010c).

References

Yan JJ, Xu F, Higano Y (2010c) Comprehensive evaluation of integrated pollutant-minimization policies in rural area around Beijing: case study of Miyun County. J Hum Environ Symbiosis 17:87–98

Yan JJ, Sha JH, Chu X, Xu F, Higano Y (2014) Endogenous derivation of optimal environmental policies for proper treatment of stockbreeding wastes in the upstream region of the Miyun Reservoir, Beijing. Pap Reg Sci 93:477–500

Chapter 7
Conclusions

7.1 Summary of Findings

The topic in this research is mainly to develop a mathematical model on the bases of some assumptions integrated with the ecosystem and economic structure based on available data to find out current environmental and economic state and future changes, which are very important to be able to formulate and introduce an optimal policy to achieve an improvement on environment and effective utilization of stockbreeding biomass resources with considerable economic growth and finally realize effective utilization of wastes, reduction of environmental pollutants and economic development. Considering the characteristics and requirements of China's economy and society, we figure out the summary of results as follows.

7.1.1 Current Situations

In Chap. 2 we calculated and analyzed the current situations of the study area from aspects of economy and water environmental pollution. According to the results of statistics, stockbreeding wastes in the study area contribute heavily to water pollution, especially pig farming has become the source of the most serious pollution in the Miyun Reservoir. The net load of T-N, T-P and COD of pig farming industry accounted about 34, 45 and 47 % of the total net load of production respectively. In addition, with the implementation of "Strong Stockbreeding County" policy, it is can be expected that the pollutants emitted by pig farming wastes would dramatically increase in the future. Therefore, we should raise and carry out proper treatment method and integrated policies to realize the simultaneous pursuit of environmental preservation and economic development in the study area. Therefore, it is absolutely of necessity and possibility to research the optimal policies and introduce advanced technologies to realize the effective utilization of stockbreeding wastes, environmental preservation and regional economic development.

© The Author(s) 2015
J. Yan, *Comprehensive Evaluation of Effective Biomass Resource Utilization and Optimal Environmental Policies*, SpringerBriefs in Economics,
DOI 10.1007/978-3-662-44454-2_7

7.1.2 Evaluation of Water Pollutants-Minimization Policies with Present Technologies

In Chap. 3, we adopted the integrated policies to minimize water pollutants flowing into Miyun Reservoir based on two sub-models. With strict restrictions on the objective function to minimize T-N, the annual growth rate of GRP only reached 1.1 % with the global solution and no feasible solution could be obtained for greater than 1.1 %.

We defined the simulation cases that reflected the level of economic development (gross regional product, GRP) as increases of n % per year (annual growth rate of GRP) during the future 10 years in the study area. Four representative cases were selected to demonstrate the variations in the simulation results. These cases were numbered so as to facilitate the simulation analysis. When a GRP increase of 0 % per year, we set the name as Case 1 to represent the minimum level of economic development. When GRP realize an increase of 0.7 % per year, we set the name as Case 2. And Case 3 shows an increase of 1 % per year. Case 2 and Case 3 were selected to express the tendency of simulation results. When a GRP increase of 1.1 % per year, we set the name as Case 4 to indicate the maximum level of economic development in this simulation.

When introduction of the optimal pollutant-minimization policies and GRP increased by 1.1 % per year (Case 4), the objective value of T-N (sum of T-N for 10 years) decreased about 10.7 % as compared to when not adopted. The feasible optimal solution was achieved in case 4 when we adopted the integrated polices. Moreover, with the integrated policies, the GRP had a difference of about 192.5 million RMB as compared to the initial year.

The following results were obtained by this simulation. First, when the integrated pollutant-minimization policies with wet methane fermentation technology were introduced, GRP increased 1.1 % per year, and the objective value of T-N decreased about 10.7 % as compared to when the policies were not adopted. Second, with the integrated policies, GRP had a difference of about 192.5 million RMB as compared to the initial year. Third, we found that 266.7 tons of T-N, 105.1 tons of T-P and 1,641.1 tons of COD emitted by the pig farming industry can be reduced by introducing the integrated policies.

Based on the results of the simulation, the adoption of wet methane fermentation technology reduced a fraction of water pollutants emitted by the pig farming industry in the catchment area. However, the limited environmental preservation is possible at the cost of significant reduction of the production of pig farming industry (the production decreases 30 %) and slow-growing regional economy (annual growth rate of economy only reach 1.1 %). In addition, there is security problem of the sealing device in the application of wet methane fermentation technology in China. On the other hand, with increased demand for animal products in Beijing and the development of a "Strong Stockbreeding County" by the Miyun government, there is no reason to think that we must deteriorate the socio-economic activity level to improve the quality of environment. Therefore, the efficiency of wet methane fermentation technology is not sufficient to treat the high concentrations of wastes in the water and meet the requirements of future economic development in the catchment area. Especially, the pig farming industry requires the introduction of advanced technology to allow simultaneous pursuit of environmental preservation and economic development.

7.1.3 Comprehensive Evaluation of Environmental Policies with Advanced Technologies for Effective Utilization of Biomass Resource

In Chap. 5 we explored synthetic considerations for ecologic system evaluation with reducing the abundance of greenhouse gases (GHGs) over the whole ecosystem by evaluating changes in material balances and consider the comprehensive use of livestock feces and urine as biomass resources. We introduced two advanced technologies based on the characteristics of rural areas of big city in China and research the arrangement of the advanced technologies to realize effective utilization of stockbreeding biomass resource.

When we adopted the policy to introduce two advanced technologies, the policy was a very effective tool to reduce environmental pollutants in all simulations. The introduction of two advanced technologies raised the level of economic growth by 10 % as compared to not adopting the advanced technologies policy. When the two advanced technologies were introduced in Case 10, the objective value (total T-N) showed a reduction of about 13 % as compared to not adopting the advanced technologies. Moreover, we also found that if the policy to introduce the advanced technologies was adopted in the catchment area, production of the pig farming industry increased about 448 million RMB Yuan over the total ten year simulation period as compared with not adopting.

Concerning environmental pollutants, we found a reduction of about 19.7 % (168 thousand tons) of greenhouse gases emitted by the pig farming industry in the total ten year simulation (2007–2016) when we introduced the advanced technologies in Case 20. In addition, we found that about 47 % (6,424 tons) of T-N, 53 % (2,716 tons) of T-P and 54 % (42,729 tons) of COD emitted by the pig farming industry in the total ten year simulation can be reduced by introducing the advanced technologies.

Moreover, from the aspect of greenhouse gases, the advanced technologies emit much less CH_4 and N_2O than any other method of animal waste treatment. This indicates that utilization of biomass resources contributes to mitigation of the GHG, even if CO_2, a generated biomass resource is not considered as carbon neutral. In this simulation, about 17 thousand tons of greenhouse gas can be reduced in the total ten years by introducing the two advanced technologies. From the aspect of biomass electric energy, more than 32.7 million kwh of electric energy can be generated when we adopted the two advanced technologies.

From the aspects of different regions, although Chao River basin has smaller amount of pig, its large-scale pig farms are concentrated, more *New Energy Project* should be installed in Region 1; since the distribution of pig farms is dispersed in Bai River basin, more *Biomass Plant* should be installed in Region 2.

7.2 Conclusions

Finally, conclusions with policy suggestions are introduced to make it possible to contribute as an appropriate action expressed briefly in the following.

7.2.1 The Modeling Approach to Address the Problem

According to the results, it can be concluded that the integrated approach is an effective method in providing valuable information that are linked to the sources of pollution, to their contributions and relation with economic indicators in the rural areas around big cities in China and by the purpose of introducing policy instruments to address the environmental problems. In addition, this study indicated that any approach to dealing the environmental problems can not succeed if socio-economic and environmental status and their interrelation are not taken into consideration based on the characteristics of the study area.

7.2.2 The Source of Pollution and Future Changes

This study has found the sources of water pollution, the contributions by each industry and the impact of future socio-economic changes on the environment in the catchment area. The main sources of pollutants are stockbreeding industry wastewater, especially pig farming industry wastewater and other stockbreeding industry.

As a result, we would emphasize the current socio-economic policy with the adoption of general wet methane fermentation technology is not sufficient in reducing the pollutants in the future. The reasons are represents as follows.

1. The pig farming industry has the largest water pollutants emission coefficient, especially T-N.
2. With the requirements of Miyun Reservoir and objective function, serious generation of pollution restricts the development of the pig farming industry.
3. In addition, there is security problem of the sealing device in the application of wet methane fermentation technology in China.
4. Miyun government has raised a strategy for "Strong Stockbreeding Industry" as a regional development orientation in order to improve regional economic development and satisfy the increased demand for animal products in Beijing. It is impossible to reduce the pig farming production in reality.

This result shows that the contribution to water pollution by wastewater from the pig farming industry is higher than considered and the efficiency of wet methane fermentation technology is not sufficient to treat the high concentrations of wastes in the water. Therefore, it is necessary and essential to introduce advanced technologies to treat the serious pollution problems and improve regional economy. The most important factor for the development of advanced technologies in the future is to enable treatment of wastewater from pig farming for higher concentrations of pollutants.

7.2.3 Cost-Benefit Options for Reduction of Pollutants and Utilization of Biomass Resource

According to the comprehensive evaluation results, the introduction of two advanced technologies for pig farming industry is very effective and the utilization of biomass resources allows simultaneous pursuit of environmental preservation and economic development. Therefore, it is essential in determining the socio cost-benefit option of two advanced technologies to address environmental pollution issues in the study area that would be feasible as an appropriate action.

The private cost-benefit analysis results represent that *New Energy Project* is more feasible and acceptable as compared to *Biomass Recycle Plant* from the aspect of private financial risk. When the subsidization rate for the technology of *Biomass Plant* was increased to 47 %, the pig farm or enterprise that introduces *Biomass Plant* technology would be able to break even and start making profits.

The simulation carried out in Chap. 6 shows that 151 of *Biomass Recycle Plants* and 14 of *New Energy Projects* should be installed to treat the pig wastes. The total cost is about 1.09 billion RMB Yuan over the total ten years. At the same time, the accumulated total GRP would increase about 1.82 billion RMB Yuan over the same period.

We also calculated the added value of advanced technologies as four parts, including biomass electric value, organic fertilizer value, CO_2 mitigation value and reduction of environmental fines. Therefore, these results of cost-benefit analysis show that there is great feasibility and possibility to introduce the two advanced technologies in the catchment area.

In the rural areas of China, central government and local government provide subsidy for the promotion of environmental preservation and effective utilization of biomass resource. However, some of the local governments have not sufficient subsidy and financial funds to provide for the introduction of advanced technologies. Therefore, when we introduce advanced technologies to utilize biomass resource, preferential policies becomes essential, the government should implement preferential policies, such as carrying out tax preference and raising subsidized rate to encourage the utilization of advanced technologies.

7.2.4 Policy Proposals for Comprehensive Management

In this research, we raised and analyzed the integrated policies for the study area to realize the simultaneous development of environmental preservation, biomass resource utilization and regional economy. It should be emphasized that modeling of ecosystem integrated with environment is one of the key points to draw the whole system in rural areas around big cities in China that we have considered as a first step of comprehensive action. In order to integrate all the systems, water pollution indicators and green house gases indicators originated by household, production, stockbreeding industry and land use activities should be classified, analyzed and collected as data base. Second, main socio-economic indicators are also categorized and determined in rural areas to be able to formulate and integrate the systems into a modeling approach.

Beside these, we should notice that without region management approach, the environmental pollution issues can not be addressed appropriately. In each region in the catchment area, local and central government with all private sectors should be organized and represented into an organization to establish and implement an action plan considering socio-economic state, sectors demands and priorities related to the environmental pollution.

This study also indicated that the synthetic policies for the catchment area, especially the introduction of advanced technologies for pig farming industry are very effective and the utilization of biomass resources allows simultaneous pursuit of environmental preservation and economic development. However, some of the local governments have not sufficient subsidy and financial funds to provide for the introduction of advanced technologies, therefore, when we introduce advanced technologies to utilize biomass resource, attraction of investment from nongovernmental sources and other private capital, including the loans from the banks are feasible options. In addition, the government should implement preferential policies, such as carrying out tax preference and raising subsidized rate to encourage the utilization of advanced technologies.

Finally, it worth mentioning that this comprehensive evaluation is expected to improve the biomass resource utilization and environmental reservation and form the basis of policy decision-making for sustainable development of rural areas of big cities in China.

This study is a first attempt to give a comprehensive analysis of environmental pollution and biomass resource utilization issue in rural areas around big cities in China. According to the basic simulation system and model of Japan, we improved the simulation model and considered the specific and special characteristics of China's economy and social state in terms of sustained economic growth rate, financial subordination relations and regional environmental policies which are different with the model of Japan. Moreover, we introduced two different advanced technologies from Japan and China to the study area through simulation with integrated policies and carried out regional analysis and allocation for two technologies which show strong operability in practice under the condition of limited funds and current states in China.

During the research we observed that it is very handy in providing a systematic assessment of the problem and integrated applicable policy. The quantitative approach employed in this study is reasonable and gives comprehensive and scientific results. It offers researchers a benchmark for applying this new methodology in their researches. The estimation methods as well as the construction of the model are extensively described. The type of qualitative information and quantitative data are given. The approach used in this study can, therefore, be utilized to address similar problem found in other regions. The policy recommendations can work as an agenda for the government. This is particularly applicable since the study suggested a certain treatment system and the application of advanced technologies specifying the required financial sources.

References

http://www.chinatouristmaps.com/travel/beijing/city/topography.html. 9 Jan 2009

Yan JJ, Sha JH, Chu X, Xu F, Higano Y (2014) Endogenous derivation of optimal environmental policies for proper treatment of stockbreeding wastes in the upstream region of the Miyun Reservoir, Beijing. Papers in regional science, vol 93, pp 477–500

Appendix
Meaning of Technical Words Used

Abbreviations

COD	Chemical Oxygen Demand
CH_4	Chemical Formula of Methane
CO_2	Chemical Formula of Carbon Dioxide
GHG	Green House Gases
GRP	Gross Regional Product
I–O	Input–Output
IPCC	Intergovernmental Panel for Climate Change
kWh	Kilowatt-hour
NH_3	Chemical Formula of Ammonia
N_2O	Chemical Formula of Nitrous Oxide
OECD	Organization for Economic Co-operation and Development
RMB	Currency of China
T-N	Total Nitrogen
T-P	Total Phosphorous

© The Author(s) 2015
J. Yan, *Comprehensive Evaluation of Effective Biomass Resource Utilization
and Optimal Environmental Policies*, SpringerBriefs in Economics,
DOI 10.1007/978-3-662-44454-2